GROWING WINTER FOOD

Growing Winter Food

CompanionHouse Books™ is an imprint of Fox Chapel Publishers International Ltd.

CompanionHouse Project Team
Vice President–Content: Christopher Reggio
Editor: Amy Deputato
Copy Editor: Katie Ocasio
Design: Mary Ann Kahn
Index: Elizabeth Walker

ISBN 978-1-62008-326-0

Library of Congress Cataloging-in-Publication Data
Names: Gray, Linda Pamela, author.
Title: Growing winter food : how to grow, harvest, store, and use produce for the winter months / by Linda Gray.
Description: Mount Joy, PA : Fox Chapel Publishing, [2019] | Originally titled: Grow your own winter food, first published in the United Kingdom in 2011 by New Holland Publishers. | Includes index.
Identifiers: LCCN 2018050249 (print) | LCCN 2018057679 (ebook) | ISBN 9781620083277 (ebook) | ISBN 9781620083260 (softcover)
Subjects: LCSH: Food crops. | Fruit-culture. | Vegetable gardening. | Herbs. | Winter.
Classification: LCC SB175 (ebook) | LCC SB175 .G738 2019 (print) | DDC 635--dc23
LC record available at https://lccn.loc.gov/2018050249

Fox Chapel Publishing
903 Square Street
Mount Joy, PA 17552

Fox Chapel Publishers International Ltd.
7 Danefield Road, Selsey (Chichester)
West Sussex PO20 9DA, U.K.

www.facebook.com/companionhousebooks

We are always looking for talented authors. To submit an idea, please send a brief inquiry to acquisitions@foxchapelpublishing.com.

Printed and bound in Singapore
22 21 20 19 2 4 6 8 10 9 7 5 3 1

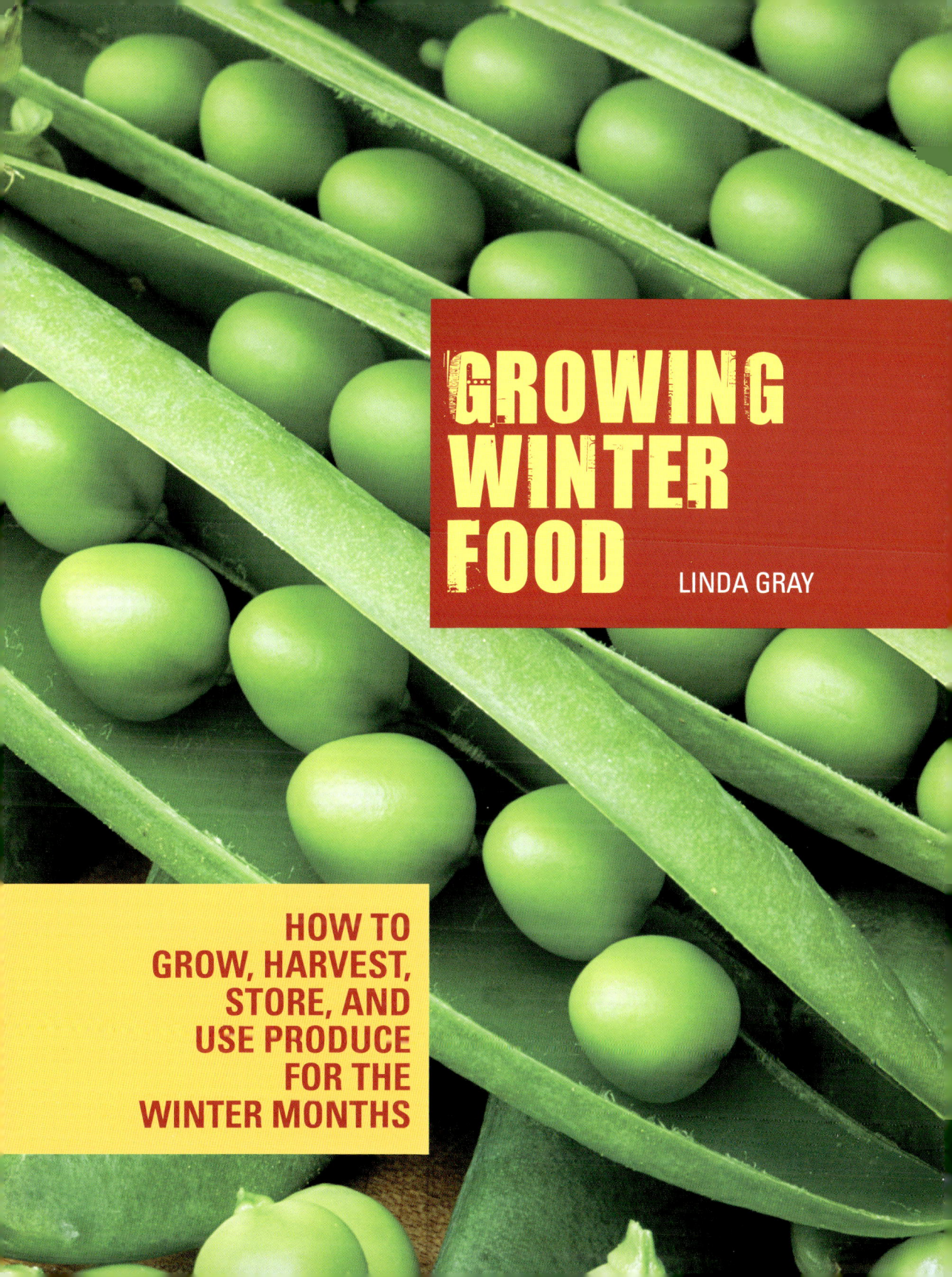

GROWING WINTER FOOD

LINDA GRAY

HOW TO GROW, HARVEST, STORE, AND USE PRODUCE FOR THE WINTER MONTHS

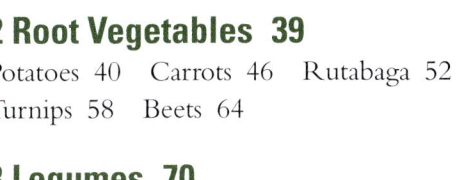

CONTENTS

INTRODUCTION

A couple of years into my adventure with an acre of land in rural France, I realized we were eating food from the garden right through the winter months as well as all of the salads and fruits during the summer. On top of the financial benefits of growing our own food, the bonus to this year-round production was the effect it had on our health.

What could be better than getting through a winter without one sniffle, cold, or any other bug that circulates every time the temperature dips? When you grow your own food, you are not only eating the best possible food on the planet, you are also getting plenty of fresh air and exercise without having to go to the gym.

This book aims to help you through the sometimes overwhelming gardening experience with step-by-step instructions on how to grow your own food and the best ways to store it. There are also recipe ideas to help you make the very best of your crops.

Read through the gardening tips and helpful advice at the beginning of the book, especially if you've never attempted growing your own food before. But even for seasoned gardeners, there are always useful tips to pick up. At-a-glance tables provide quick references to sowing and harvesting times, pests and problems, and storing recommendations.

The second chapter has detailed instructions for growing a number of everyday root crops with traditional advice on how to store them. Some can even be left in the ground right through the winter months. In the third chapter are legumes; although peas and beans are not generally grown during the winter months, they are exceptionally easy to grow and will store almost indefinitely, particularly if dried.

All of the green vegetables listed in the fourth chapter grow happily through the colder months of the year, and some even improve in taste after being frozen on the plant a couple of times. Herbs are a must-have if you want to get the most from your crops, and they are also very accommodating plants to grow, as you will see in the fifth chapter. Many can be grown indoors in containers or outside in specially prepared herb beds. Herbs are also great companion plants to grow in your vegetable patch to deter pests and viruses from damaging your valuable fruit and vegetable crops.

The last chapter focuses on fruits. Usually considered summer-only crops, fruits are now available in a surprising number of hybrids that will crop right through the autumn and early winter. Those that don't will store well in a variety of different ways.

Not only is gardening a creative pastime that gets you moving and outside in the fresh air, it also puts you in touch with the earth, which is just as important. The soil contains natural antibiotics, so a regular dose of nature combined with eating the best organic food on the planet will boost your immune system and help you stay healthy all year.

Enjoy your gardening experience!

Pears are among the fruits that can be grown and harvested for winter eating.

1 GENERAL GARDENING TIPS

If you are facing a mini-jungle behind your house or simply don't know where to start, the best way forward is in very small steps. It's easy to become daunted by the task at hand, and one of the most common causes of back complaints—especially after the first sunny weekend of the year—is overzealous digging and mowing.

Take it slowly. The garden is an ongoing project and should never really be "finished." Plants grow, change shape, are replaced, and need a certain amount of maintenance to keep them healthy. It is possible to buy an "instant" garden these days, but it will never have the same appeal or produce the same kinds of crops as a traditional garden that has been lovingly created.

Enjoy Your Garden

Where do you start? The best way to get the most from your outdoor space is to make a plan. Putting your ideas and thoughts on paper will make the way forward a lot clearer, and you can tackle one part of your plan at a time.

Spend a little time in the garden before you start. Looking out the window won't give you a full picture, no matter how wonderful your patio doors are. Get out there, even if it's just for a few minutes every day, to observe and answer a few important questions:

• Where are the sunniest spots in the garden?
• What parts of the garden does the sun never reach?
• Is the soil well drained? Are there any waterlogged areas? Does the water sit in one spot?

Decide where your fruits and vegetables will grow. Pick the sunniest place if you can. Although some plants prefer a bit of shade during the day, most will benefit from a sunny spot. Make sure that the soil is well drained in this area because few plants will grow successfully in too much water (one exception to this rule is watercress).

Check the pH balance of the soil. You can easily find a pH-testing kit at most large garden centers or via garden suppliers. It's useful to know whether your soil is on the acidic side because different species of plant will thrive in different soil types. Brassicas (cabbage, broccoli, rutabaga) won't thrive and develop well in acidic soil. If your soil is out of balance for what you want to grow, adjust it long before you start sowing seeds or placing plants. Dig in some lime or other organic material made especially for the job and let it settle. This is probably best done either in early spring, as soon as the soil is workable, or at the end of summer in preparation for spring planting.

Think about all of your garden needs. If you have children, a play area is probably a good idea unless they are all enthusiastic gardeners. Include pathways, however simple, to allow for movement, access, and easy maintenance. And, most importantly, create a space for relaxing. Whether it's a comfy chair on the patio or a garden seat around the apple tree, you should always try to incorporate a seating area in your garden plan. A place to unwind and think about the next garden activity is an absolute must.

Ideas

There are many ways to get the most out of your garden space, and it's easy to get carried away with the choices available. Try to keep to a modest plan and build up slowly if you are just beginning to garden. Even if you have to hand a whole area over to nature for a year or two, it is not the end of the world. In fact, a wild area of the garden will attract bees, butterflies, and maybe even a few friendly frogs, which are always welcome garden guests. Amphibians will keep the slug population down, which in turn could save many of your young plants. Slugs and snails are the gardener's

nightmare because they can devour a whole line of newly germinated plants in a single sitting.

Bees will help pollinate the plants all around the garden, and, of course, butterflies are always welcome visitors, too. Having said that, you should watch the cabbage white butterfly closely. Once the cabbage white lays her eggs on the underside of your cabbage leaves, the caterpillars that hatch after a few days can devour the whole plant with the same ease and speed as slugs can destroy lettuce plants.

To get started in your garden, don't worry about getting everything all done all at once. Introducing even one new plant every season or every year is progress however you look at it. It's often said that you should live in a house for twelve months before you start making changes to it. The same could be said of the garden. It's good to know what you are dealing with before you start. And your garden should be an extension of your home, not an extra chore. When you make plans and allow yourself enough time to do it, gardening can become a relaxing, healthy, and inexpensive hobby.

Vegetables

There's nothing more rewarding than growing vegetables in your garden for family and friends to enjoy. Although there are initial expenses and a certain amount of "work" involved, when you pick fresh produce straight from your own vegetable patch, it always feels like free food.

Your vegetable patch should generally be in a bright, airy spot that is sheltered from the wind and preferably in a sunny position.

Make sure you know which area gets the most sun throughout the day.

You could dig over a large square in which to grow your vegetables or go for smaller beds. "Potagers" were invented by the French to grow more crops in a smaller space without adding to the workload. If you want to recreate this idea, make small, square beds and plan to grow vegetables in the middle and herbs around the edges. However, this is your space, and you should grow what you feel is best for your region, your garden area, and your family.

Root crops need depth of soil, so if you have very shallow soil in your garden, you may have to create raised beds and add some topsoil. Alternatively, you could build up the soil yourself over a few years, using soil from your own compost heap.

Acquire some well-rotted manure if the soil seems to be tired and lacking in nutrients. When you get in touch with your garden, you will be able to tell instinctively if the soil is healthy and rich enough to feed your fruit and vegetable crops. If it seems a little on the poor side, dig in some well-rotted manure or compost to add nutrients to the soil before you start. You should not add fresh manure to the soil just before planting because it will be too strong for very young plants. Dig it in at least a month or so before planting or spread manure at the beginning of autumn.

When you are sure that the soil is healthy and deep enough to accommodate your crops, dig the area over when it's not too wet. Don't try and dig over soil that is very

wet and heavy. It not only hurts your back, but it can also change the consistency of the soil, causing it to clump together and reduce its ability to release nutrients for your plants to take up.

Every crop you grow will draw nutrients from the soil, and you should replenish the soil as often as you feel necessary, but at least once every couple of years if you want to grow healthy crops. Make feeding the soil a regular habit by spreading organic fertilizers over it or digging them into the soil as soon as you have cleared a space.

Plan to rotate your crops every year. Planting the same crops in the same place will encourage diseases and draw the same nutrients from the soil. Plants in the

An apple tree, heavily laden with ripe fruit, is touched with a gilding of autumnal frost.

same family also should not be planted in the same spot year after year either. For example, potatoes, tomatoes, and peppers all belong to the same group of plants and can pass on unfriendly diseases through the soil. Rotating crops helps prevent these problems, and it's worth breaking your vegetable-growing areas up into three or four working spaces.

Fruit

You don't need an orchard-sized garden to grow fruit these days. There are many miniature fruit trees available from good garden suppliers. Always make sure that you check the manufacturer's growing instructions and recommendations before

you buy so you know that you have the right space for your plant. There's no point in putting a sun-loving peach tree in a shady place in your garden. You may be able to keep the tree alive, but it's doubtful that it will ever produce a good crop of fruit.

Native trees are always a good starting point if you want to grow fruit or other trees in your garden. These tend to be more resistant to local bugs and viruses and will cope with the weather conditions far better than plants originating from other climates.

Larger-growing trees often need little or no maintenance and will keep cropping year after year, although you must take care when planting them. Position a larger tree carefully, especially if it is likely to grow to a whopping 30 feet (9 m) high.

There should be pruning instructions provided with the tree when you buy it. Some fruit trees can be trained to fan out over a fence or a wall. These fruit trees are generally grafted onto sturdy rootstock and can be resistant to many diseases.

As well as fruit trees, you can also incorporate soft fruits, shrubs, and perennial plants into your garden. Strawberry beds are fairly easy to maintain and will thrive year after year with a little care and attention, rewarding you with the best-tasting strawberries you have ever eaten. Likewise, raspberry and blackberry canes are worth planting for their luscious fruits, which you can store for winter eating.

A windowsill herb garden can "green up" your indoor space all year round.

Herbs

Planting an herb garden really is a wonderful way to brighten up your outdoor space. The good thing about herbs is that most will grow well in containers and thus can be planted in pots on the patio or balcony. Many can even be grown on a sunny windowsill. Herbs tend to crop throughout most of the year. Some are evergreen, and others can be dried and stored for the winter months.

Plan your herb garden carefully, and it will thrive for many years to come with very little maintenance required. Herbs can be strategically placed to give off the most delicious scents when brushed against, and many also have the bonus of deterring bugs and pests from your vegetable crops. Growing a few in your vegetable patch will help protect your plants, and you will also remember to pick them.

Themes

A garden theme can take your whole garden space to a new level. Color-coordinated flowers create a spectacular focal point to your garden, as will an herb garden or a rose bed.

You can grow plants specifically to attract butterflies or bees. Make sure that you have a small water feature available for wildlife to encourage return visits to your garden. Water features in a garden not only provide excellent feng shui and are beautiful to look at, they can also be a practical addition to the space. For example, you can grow watercress in running water. Frogs and toads will be attracted to your garden if you have a pond available for them.

Even if you don't have a large garden area, a patio can become a thing of beauty rather than just an extra surface to sweep from time to time. Grow herbs and flowering shrubs in attractive containers. You can pick up old pots at thrift shops or yard sales and, with a little creative flair, you can turn them into visually appealing containers for your patio.

Make a Plan

Once you have decided what you want from your garden space, draw up a plan. It does not necessarily have to be to scale, although you will get a better idea of what you need if you take the time to measure the garden and add a little precision to your drawing.

From the plan, you can figure out what materials, tools, and equipment you need for each part of the garden. Again, especially with a large garden, don't try to take on too much all at once unless you have an army of willing helpers and an unlimited budget.

Decide what you will be working on in the coming year and plan accordingly. Start growing food crops as soon as you can. Even if you've dug out only a very small area, producing your own food from the start will encourage everyone in the family to join in with the preparations the following year.

Keep a garden journal and make note of when you sow seeds or transplant plants. Later, you can jot down how each crop fared and use this information to either repeat a good crop or make changes if the results were a little disappointing.

Tools and Equipment

Because you need to keep your garden tools and equipment in a dry space, the best thing you can have in your garden is a shed. But before you buy or build a shed, consider what you will be storing in it. If your garage is large enough, you can store your garden machinery there rather than in a shed.

Store hand tools, small tools, pots, and other paraphernalia in a shed. This will protect your equipment and also help keep the garden tidy. Pots lying around are reminders of unfinished projects! Having a shed with a designated shelf for pots will encourage everyone to clean up after themselves.

Another way to store small hand tools and gardening gloves is to position a mailbox or birdhouse with a cover in the vegetable plot.

This saves you from having to search in the shed if you just want to work in the garden for a short time.

No matter how you decide to store them, tools must be kept under cover to protect them from weathering. Metal implements rust very quickly if left out in the rain and dampness. Well-cared-for tools encourage use and will last many years.

Gardening will be easier if your equipment is organized and stored in a handy location.

Greenhouses

A greenhouse is definitely an advantage if you intend to start your plants from seed, and some plants can be kept in the greenhouse to grow. Choose a greenhouse according to your needs and budget. There are mini-greenhouses available from good garden centers and online suppliers, and one of these will give your seeds a good start while not taking up too much room in your yard. You could also place a smaller greenhouse on a patio.

Cold Frames

Like greenhouses, cold frames are a distinct advantage in cooler climates, providing much-needed protection for tender plants.

You can put cold frames together with a few sheets of glass and a little carpentry expertise, or you can buy some from a local garden supplier. Cold frames can be positioned next to a greenhouse or on the edges of a vegetable garden. Position them carefully to avoid accidents and harm to children and pets.

Cloches

There are different types of cloche. One type is simply a miniature hoophouse. Make one by bending a long length of plastic tubing or similar material into an arch and pushing each end firmly into the ground; do the same with several more pieces of tubing in a row to form a tunnel shape. Make sure

Brick and glass cold frames. Much smaller versions are available to fit different budgets and space requirements.

A bed of lettuce seedlings under a mini-hoophouse cloche.

that there are no rough edges. Cover the arches with a clear plastic sheet and weigh the plastic down along each length with bricks or logs to hold it in place.

Another type of cloche is a covering for an individual plant, like a glass jar or the bottom half of a plastic bottle. You can put these portable cloches over young plants at night if frost is expected and then remove them the next morning—they function almost like an electric blanket for tender plants. Again, cloches of all shapes and sizes can be found at garden suppliers or even made with materials you already have at home.

Tools

Depending on what garden projects you want to tackle, you will need a variety of tools. If you will be regularly cutting lawn areas or trimming hedges, consider the following:

• Lawn mower
• Edge cutter
• Hedge trimmer
• Pair of strong, sharp shears

Tackling a small area at a time will require hand tools, and you will eventually need

Purchase quality hand tools that will last for many seasons in the garden.

them all over the garden, so consider your purchases carefully. Don't skimp on quality. Cheap tools can bend, buckle, and break easily and will have to be replaced. Weak-handled or badly joined tools can cause accidents. The extra investment in quality tools is worthwhile. You do not have to buy every tool in the garden center to get started, so start off with the following basics:

• Sharp-edged spade

• Garden fork

• Hoe

• Rake

   Make sure you know how tools, large and small, work before you buy them. Hold them or handle them in the store and make sure that they are comfortable for you. For example, a spade with a handle that is too short will cause unnecessary back strain, and so will anything that is too heavy. If you can hardly lift it, you won't be able to use it in the garden very well. Buy sensibly for your own strength and needs.

Other Tools and Equipment

There is such a range of garden paraphernalia available to buy and, because it's so much fun to wander around a garden center, it's easy to fill the shed with stuff you don't really need and probably won't ever use. Here are some tool-buying tips:

• Be practical and buy what you need to start your garden, adding more tools and equipment as and when you need them.

• Buy a strong hand trowel that feels comfortable in your hand.

• A pair of good-quality, sharp pruning shears is a wise investment for pruning fruit bushes, roses, or other small shrubs and plants.

• Good-quality, strong gardening gloves are invaluable to protect your hands against stinging nettles, sharp thorns, and staining of the skin. If you have any plant allergies, you must wear gloves. Try them on before you buy them.

• There are "bionic" styles of gardening gloves available, although you may find that they are not flexible enough for you. Again, try them on before purchasing.

• However tempting it is to stroll outside in your slippers on a sunny day, they are not practical footwear for the garden. Wear a sturdy pair of boots to protect your feet when digging or using other sharp tools. Boots also give you some stability when using tools with which you may not be familiar.

   If you are planning to sow seeds, you will need seed trays and individual pots according to the needs of your particular seeds. Root crops are generally always sown directly outside, so if you are planning to grow only a few roots this year, you probably won't need seed trays or pots at all. However, if you have trays and pots available, you can use them to try your hand at a few different crops. You will start most summer vegetables off in early spring in trays or pots and then keep them warm and watered until you transplant them into the garden later.

Buy fairly sturdy pots and trays so you can use them more than once; flimsy ones are awkward to use and will break easily. Also consider biodegradable pots. With a biodegradable pot, you plant the whole pot into the soil when the plant is ready to be put outside. This way, you won't disturb the plant, which prevents damage to the roots. Root crops can fork or split when transplanted, but you will avoid this risk if you use individual biodegradable pots. See the individual growing advice later in the book to find out which roots can be started off this way.

Free Pots and Seed Trays

• You can make biodegradable pots with a pot-making kit, available from good garden suppliers, or by simply rolling a couple of layers of newspaper around something like a rolling pin and then folding under the bottom edge to form a pot. Fit them closely into a seed tray before filling them with soil. They won't last long but should make it until you need to plant them out in the garden.

• The cardboard tubes from paper-towel or toilet-paper rolls can also be used as biodegradable pots, so make a point of saving them. Cut the longer tubes in half, and you have two for the price of one.

• Plastic yogurt cups are fine for small plants. Be sure to wash, rinse, and dry them well and then punch a couple of drainage holes in the bottom of each one.

• Paper egg cartons and trays make ideal biodegradable pots for individual plants.

Planting in biodegradable pots allows you to transplant without disturbing the seedlings' delicate root structure.

These rustic DIY pot markers were made with wood scraps and pieces of fallen tree branches.

• Old paint trays are good alternatives to seed trays. Wash and rinse them well, dry thoroughly, and punch some drainage holes in the bottom.

• Line shallow wooden fruit and vegetable crates with cardboard and use as seed trays.

Other Recycled Supplies

• Keep old ice-pop sticks to use as pot markers. The wooden ones are easy to write on with a pen or pencil—no special marker required.

• Save used wooden skewers to mark row ends in the seed bed or garden.

• Clear plastic sheets are invaluable in the garden to protect young plants from cold nights and to give your seeds an extra boost in the greenhouse.

• Check packaging material before you throw any of it away because some of it might be handy in the garden.

• You can bend old wire coat hangers into arches to use as frames for clear plastic to cover your smaller plants.

• Cut clear plastic water bottles in half and use them to protect individual plants on cold nights.

• Old CDs or aluminum foil strung across a bush or bed will frighten the birds away from your crops.

• Use wire fruit and vegetable baskets as hanging baskets. Hang one outside the kitchen door with a few herbs growing in it.

• Recycle household items such as cutlery, pots and pans, cups, bowls, and teapots to use in the garden. It doesn't matter if they last for only a season if they are free.

Free Soil

Produce your own free organic soil by building a compost heap. You can buy a composting tumbler if space is limited; otherwise, build one yourself. Ideally, the compost heap should be accessible and easy to reach from the house so it won't be a chore to take the vegetable peelings out.

Don't build the compost heap in a sunny spot. A shady part of the garden will be perfect. Make sure that the heap has adequate airflow so it can breathe and break down the organic matter. A compost bin made of wooden slats positioned in a square around the sides of four posts works well. The bottom slats of one side should be removable so you can dig out the fresh compost from the bottom of the heap when it's ready. Spread fresh compost over the vegetable plot in the autumn months to nourish the soil, or dig it into the ground a month or two before planting your crops.

Free Mulch

You can compost grass clippings, but grass clippings are also ideal to use as mulch around your plants. Spread clippings around lettuces and cabbages to keep their bottom leaves clean and the weeds away. Water the mulch after spreading to flatten it. Don't mulch too high or too close around your plants, or the stems may rot.

A backyard compost heap yields nutrient-rich soil for your plants.

COMPOST TIP

Always use fresh compost rather than garden soil for your pots or trays of seeds and seedlings. When transplanting, mix the compost from the pots with the garden soil and plant your new plants in this mix.

You may use pine needles and other hedge clippings, but because they tend to be more acidic in content, only use them to mulch around the plants that like acidic soil.

Seasonal Sowing and Harvesting

Although we tend to associate sowing and harvesting with spring and autumn, respectively, many crops can be sown and harvested at other times. The following charts show at a glance the sowing and harvesting times for all of the crops in this book. While you can refer to the given seasonal guidelines, it is always best to go by the weather, your knowledge of your own climate, and your instincts. Also check the instructions on your seed packets for regional variations.

Crop	Sowing/Planting	Harvesting
Potatoes	Late summer (with plastic or glass cover toward end of year)	Late winter onward
	Spring (main crop)	Summer–early autumn
	Midwinter (under cloche)	Spring
Carrots	Late winter–early spring (under cloche)	Summer
	Spring–summer	Summer–autumn
Rutabaga	Late spring–early summer	Early autumn–early spring
Turnips	Early spring (summer varieties)	Midsummer
	Summer (main crop)	Early winter–midwinter
Beets	Spring–summer	Summer–autumn
Green beans	Late spring–early summer (French beans)	Summer–autumn
	Late spring–early summer (runner beans)	Summer–early autumn
Fava beans	Spring	Summer
	Late summer–early autumn	Early spring
Peas	Spring–summer	Summer–autumn
Kale	Late summer–autumn	Winter–early spring
Brussels sprouts	Spring	Winter–early spring
Winter lettuce	Late summer–autumn	Autumn–spring

Crop	Sowing/Planting	Harvesting
Sage	Spring (seed)	From following spring
	Late summer–autumn (cuttings, layering)	From following summer All year round when established
Rosemary	Spring (indoor seed)	From following spring
	Early summer (outdoor seed)	From following summer
	Late summer–autumn (cuttings, layering)	From following summer–autumn All year round when established
Thyme	Spring (indoor seed)	From following spring
	Early summer (outdoor seed)	From following summer
	Late summer–autumn (cuttings, layering)	From following summer–autumn All year round when established
Garlic	Early spring	Summer–autumn
	Autumn	Spring
Chives	Early spring (seed)	From following autumn
	After flowering (root division)	When plants have settled and are growing again
		All year round when established
Parsley	Early spring (seed)	Summer–late autumn
Apples and pears	Autumn–winter (one- or two-year-old trees)	Late summer–early autumn
Strawberries	Early spring (seed)	Following summer
	Late summer–early autumn (plants)	Early summer, some varieties until late autumn
Blackberries	Late summer–autumn	Summer–autumn
Black currants	Autumn–spring	Summer–autumn

Practical Tips

There are so many variables in the garden, so the better armed you are against problems, the more successful your crops will be. The state of your soil, the weather, and a vast collection of garden pests can conspire against even the most well-informed gardener.

Learning how to protect your own plants from the elements, pests, and diseases soon becomes second nature, especially after you have eaten your own homegrown crops and discovered just how delicious and nutritious they are. The following practical tips cover many common problems and will help you get the most out of your garden regardless of its size.

The Soil and the Weather

If you don't have well-drained soil, incorporate either sand or other organic material to increase the drainage or build raised beds. The roots of your plants—in particular those plants that grow through the winter months—will rot in waterlogged soil. A touch of frost will freeze the water in the soil very quickly, and if the water doesn't

get your plants, the frost will. So make sure that your plants are in a well-drained spot.

No matter how conscientious a gardener you are, if bad weather strikes, there is very little that you can do. However, most gardeners tend to become very good at predicting the weather over the years, and with a little practice and observation, everyone can do it. Ignore the weather forecast. Instead, go outside and smell the air, look at the clouds, feel the temperature. Then imagine how your plants are coping.

Rain

Of course, we need rain, but if you have recently sown a delicate line of seeds, a sudden heavy shower could wash them away in seconds. Placing a simple cloche over the seeds before it rains will protect them (don't forget to remove it, though!). Constant rainfall may wash nutrients through the soil too quickly for your crops to take them into their roots. Feed the soil fairly frequently if there has been more than average rainfall.

On the other hand, don't assume that the spring rains will provide enough water for your newly planted trees. If you aren't getting a heavy rainfall every week or so, new trees in particular will need deep watering.

After it rains, go out into the garden and pull the weeds. It's much easier to pull up the whole plant, including the root, when the ground is soft, and it's easier on your back, too.

Sun

Most, but not all, crops benefit from a sunny spot in the garden. And although they may

DID YOU KNOW?

When you need to water the garden, do it in the morning or evening—or both. Watering during the middle of the day can be harmful to your plants, and it's also a waste of water that evaporates quickly in the sun.

love the sun, plants are vulnerable to the sun's rays and can burn in hot midday heat. A little shade will prevent the soil from drying out too quickly while protecting the plants from burning. Rig up something to shade your plants, but remember to remove it before night falls so that they can benefit from the early morning sun.

Wind

A cold, strong wind can devastate tender crops. When planting, try to avoid windy areas in the yard and always provide a support system for those plants that need it. If possible, position your shed in the windy spot to shield the garden from excessive wind. Make sure, however, that the shed doesn't throw too much shade on your crops.

Sunlight and autumn frost illuminate the beautiful structure of this brussels sprout.

Snow and Frost

Fruits and vegetables with high water content will almost dissolve before your eyes if caught in a heavy frost. However, there are a number of vegetables that will thrive quite happily through a frost and will still be healthy and available to harvest even under a blanket of snow. Some crops actually taste better after the first frost.

Protect tender crops under glass or plastic if you start them early in the year. By mid- to late spring, it is usually safe to assume that all danger of frost has passed, but it's always a good idea to check your seed packets for growing recommendations for your region before you start to sowing seeds or transplanting.

Bugs and Other Problems

Some plants are vulnerable to all sorts of pests and viruses, although many winter crops are fairly robust. Whatever you are planting, slugs and snails can be a problem, so watch out for them. They can eat through the stems of every seedling very quickly, killing off all your young plants completely.

Crop	Problem	Treatment	Prevention
Potatoes	Blight	Bordeaux mix	Don't plant in soil where potatoes, peppers, or tomatoes have grown in previous two years
Carrots	Carrot fly	No known organic treatment	Plant next to onions Cover with a fleece made for this purpose
Rutabaga	Clubroot	–	Plant in neutral rather than acidic soil
	Caterpillar	Remove by hand	Butterfly netting
Turnips	Clubroot	No known organic treatment	Plant in neutral rather than acidic soil
	Caterpillar	–	Butterfly netting
Beets	Caterpillar	Remove by hand	Plant in neutral rather than acidic soil Butterfly netting
Green beans	Black fly	Organic products available	Encourage ladybugs and grow marigolds nearby
Fava beans	Black fly	Organic products available	Autumn-sown plants are usually safe from black fly Encourage ladybugs, grow marigolds nearby
Peas	Mice	–	Coat seed in paraffin before sowing
	Caterpillar	Organic products available	Rotate crops (larva lives in soil) Cover with horticultural fleece from early summer

Crop	Problem	Treatment	Prevention
Kale	None known	–	–
Brussels sprouts	Clubroot	Organic products available	Plant in neutral soil rather than acidic
	Aphids	Organic products available	Encourage ladybugs, grow marigolds nearby
Winter lettuce	Root maggot	Organic products available	Water regularly and don't leave roots in the ground after harvesting

Crop	Problem	Treatment	Prevention
Sage	Waterlogging	None	Plant in well-drained soil
Rosemary	Not thriving	Add lime or other alkaline product	Plant in neutral soil
Thyme	Waterlogging	None	Plant in well-drained soil
Garlic	Bolting	Fold down leaves	Shade in very hot sunlight, water regularly
Chives	Onion fly	Organic products available	Don't plant near onions
Parsley	Not thriving	Feed soil	Plant in rich soil

Crop	Problem	Treatment	Prevention
Apples and pears	Aphids, canker, mildew	Cut away canker from bark with sharp knife, treat with Bordeaux mix if necessary	Keep clean area around trees Sweep up leaves, remove dead fruit, etc.
Strawberries	Aphids	Organic products available	Encourage ladybugs, grow marigolds nearby
Blackberries	Aphids	Organic products available	Encourage ladybugs, grow marigolds nearby
	Rust	–	Keep area clean
Black currants	Aphids	Organic products available	Encourage ladybugs, grow marigolds nearby
	Rust	–	Keep area clean

Slugs and Snails

There are organic slug repellents on the market, and they should be available from a good garden supplier or garden center. Traditional slug deterrents, including the following tips, can be useful as well:

• Break eggshells into fairly small pieces and spread around plants. Try not to leave any gaps where the slugs can make it to your plants; slugs will find a way if they can.

• Placing sand around the base of your plants can help, although, after a rainfall, the sand softens and doesn't bother the slugs much. The drier and grittier, the better.

• Slugs are attracted to beer. A small bowl of beer near your plants will distract them from dinner in favor of a few sips. Don't rely on this method too heavily, though.

Slugs and snails are the gardener's enemy, but there are a number of ecological ways to keep them under control.

The Good Guys

Not all insects and wildlife are harmful to your garden; in fact, in some cases, they can be quite beneficial.

• Frogs and toads not only delight children, they also keep the slug and snail population under control.

• Ladybugs eat bucketloads of aphids, protecting plants from black, green, and white fly attacks.

• Bees are an absolutely essential part of successful gardening, helping pollinate your plants. If you can, grow some flowering plants that bees love in order to encourage them to linger in your garden and pollinate your useful crops.

• Worms are a gardener's delight because they aerate and fertilize the soil.

There are a few other less conspicuous creatures that are well worth encouraging in your garden:

• Bats eat insects and can protect you from gnats and mosquitoes in the early evening.

• The praying mantis might not be a very well-known visitor to many suburban gardens, but if you see one, rest assured that it will devour harmful aphids as well as pesky mosquitoes.

Make friends with frogs. They will help you deal with slugs, snails, and other pests.

Alternative Planting Areas

The wonderful thing about plants is that they are not picky about where they live as long as they have the right amounts of light, warmth, nutrition, and water. If a sufficient area of a yard or garden is not available to you, there are other ways and means of producing your own food.

Be adventurous when it comes to containers. A display of containers or large pots grouped together with herbs or salad crops tumbling from each can be a glorious focal point on a patio or in the garden.

Containers

There are many plants that can be grown in containers, including some of the crops listed in this book.

Potatoes

Potatoes can be grown in deep containers, stacked rubber tires, or purpose-built potato barrels. You can keep all of these types of containers on a patio or balcony but don't tuck the containers away too much because the potatoes will still need a reasonable supply of light and air to thrive.

POTATO POINTER

A raised bed system keeps the garden organized and is easy to maintain, but I have found that potatoes don't crop as well if you don't earth them up.

Carrots

Trough-style containers work reasonably well for carrots. They must be deep enough to allow the carrots to develop fully.

Rutabagas, Turnips, and Beets

Try large containers for these root vegetables, although they tend to grow and develop more successfully in a vegetable garden.

Peas and Beans

Some varieties will grow well in large containers, but many need support systems, making them impractical for container growing. A large pot against a south-facing wall or fence to which a trellis or other support is affixed could work well.

Kale and Brussels Sprouts

Kale and sprouts generally should do well in containers, and they don't need protection from the cold. Save the sunny spots for other plants that need it.

Herbs

You can grow almost any herbs in pots, although the larger types, such as rosemary and lavender, are better grown in the ground because they will keep growing and developing for years.

Apples and Pears

Apples and pears generally should be planted in the ground, although you may be able to find a hybrid that is small enough and grows well in your region in containers.

Strawberries

Strawberries grow very successfully in specially designed strawberry planters, which you can keep on your patio or, in some cases, indoors in a conservatory.

Blackberries

Wild blackberries are not suitable container plants, but some hybrid types may work well for you in containers or large pots.

Black Currants

Smaller growing varieties may do well in containers, but they are generally happier in open ground.

Containers are a great option for gardeners with limited outdoor space.

Indoor Growing

Unless you live in a particularly bright and airy house, many food crops won't get enough light to thrive indoors except on a sunny windowsill. The best crops to grow indoors are herbs. A few pots of herbs on the kitchen windowsill will often last all year, and they are harder to forget about than those grown outside.

A conservatory is usually bright enough for growing fruit and vegetables. Containers of cherry tomatoes are attractive and useful,

and you could also try your hand at growing grapes or other fruits that benefit from a little extra protection.

Lemon trees in containers should have a space in the conservatory or another bright room in the house during the winter months if you live in a cool climate.

Raised Beds

For those who struggle with the physical bending involved in gardening, a raised bed system can be a huge advantage. The original concept of a raised bed was that you dig the ground over to about 3½ feet (1 m) across so that you can reach the middle from both sides. Then, once you've dug the ground deeply and incorporated well-rotted compost, you don't walk on it or dig it again until the following

With a raised bed system, the whole bed is easily accessible, which is a boon to gardeners who find it difficult to bend and reach.

year, when you add more nutrients. This prevents the soil from becoming compacted and allows you to grow more plants in a smaller space.

Build a more permanent raised bed with bricks or concrete blocks. Build two parallel walls about 12 inches (30 cm) apart and then create ends to join the walls with bricks, sheets of aluminum, or other materials. Make sure that the sides are solid so the soil doesn't move around.

Compost heaps sometimes reveal themselves as perfect "raised beds." When you see the odd zucchini plant coming up on its own, encourage it to grow!

Raised beds with solid sides will keep the soil in place.

Camouflage

Tall plants, such as sunflowers or hollyhocks, can cover an unattractive wall or fence as well as an unsightly compost heap. You could also trail climbing crops over south-facing fences or walls. Plant them close to the wall or fence and attach nails or other supports for the plants to cling on to. Soil close to a wall tends to dry out quickly, so make sure that your plants get enough water.

Storage Methods

If you've had a good harvest, it's very likely that you will need to store some of your crops. There are many ways of doing this, although the best storage method may vary for different types of crops. The following methods are suitable for homegrown crops listed in this book.

Notes: Never wash root crops before you store them. Dry them in an airy or sunny spot for a few hours. Brush off excess soil. Always handle produce carefully to avoid bruising because damaged crops don't store well. Freezing soups, pies, or other recipes that you've made with your homegrown produce is a good option for crops that don't freeze well in their natural state.

Crop	Storage Options
Potatoes	• Vegetable clamp (see page 43) or in single layers in a cool, dark place for several months • Purpose-built potato storage unit for several months • A few days in the vegetable rack
Carrots	• Vegetable clamp or in single layers in a cool, dark place for several months • Barrel of sand for several months • A few days in the vegetable rack
Rutabagas	• Best left in ground and used when needed until spring rainfall • Vegetable clamp or in single layers in a cool, dark place for several months • Produce compartment of the fridge for a couple of weeks • A few days in the vegetable rack
Turnips	• Vegetable clamp for several months • Single layers in a dark, airy place for a few weeks • Diced and frozen
Beets	• Pickled or bottled (will keep for many months) • Vegetable clamp for several months • Produce compartment of the fridge for a week or two
Green beans	• Frozen or canned
Fava beans	• Dried, frozen, or canned
Peas	• Dried, frozen, or canned

Crop	Storage Options
Kale	• Best used straight from the ground all winter • Produce compartment of the fridge for a couple of days
Brussels sprouts	• Best used straight from the ground all winter • Frozen • Salad compartment in the fridge for a few days
Winter lettuce	• Best used straight from the ground all winter • Salad compartment in the fridge for a day or two
Sage	• Dried or frozen • Some varieties are evergreen, so use throughout winter
Rosemary	• Evergreen, so use throughout winter • Dried or frozen
Thyme	• Evergreen so use throughout winter • Dried or frozen
Garlic	• A week or so in the vegetable rack • Several months in braids or laid out in boxes in a dark, airy place
Chives	• Can be evergreen • A day or two in the fridge • Frozen
Parsley	• A few days in the fridge • Dried or frozen
Apples and pears	• Wrap in newspaper or tissue, store in cardboard or wooden trays in single layers • Keep in a dry, airy place out of direct light • Dry in rings in an oven on low heat or a home food dryer
Strawberries	• Frozen, although some taste and texture is lost • Jams and preserves
Blackberries	• Frozen, although some taste and texture is lost • Jams and preserves • Cordials, wine
Black Currants	• Frozen • Jams and preserves • Cordials

2 ROOT VEGETABLES

Although most root crops are planted in the spring, you can successfully store them for winter eating, and they keep very well. Rutabagas, for instance, can be left in the ground well into late autumn. Your own organically produced vegetables will help keep the whole family healthy during the winter months, and root vegetables are especially nourishing and filling.

Generally, all root crops need soil of a good depth that has not been fed with fresh manure in the previous month or so, because fresh manure will cause root crops to split or fork. If you think your soil needs feeding, dig in fertilizers at least two or three months before planting or sowing seed in the spring. Alternatively, spread well-rotted manure over the ground during the previous autumn or winter so that the nutrients have ample time to be incorporated into the soil.

Potatoes

Strictly speaking, potatoes are tubers, not roots, but for our purposes we will consider them as a root vegetable because they grow under the ground and are a nutritious and delicious family standby.

Although most people will tell you not to grow potatoes because they are cheap to buy and hard work to grow, this is turning into a bit of a myth. For one thing, potatoes aren't particularly "cheap" anymore, and you can grow them in barrels, stacked rubber tires, or even a deep bucket if necessary, meaning that far less physical work is involved.

You can buy specially-designed potato barrels from garden centers or other garden suppliers; these are useful in small yards or for those with no yards at all. Stand a potato barrel on a patio or in a quiet corner, as long as it's a bright enough spot. You can leave the potatoes to just do their job of growing with very little input from you.

Seed

To get the most out of your potato harvest, start off with specially grown seed potatoes. Each one should be not much bigger than a golf ball and should look healthy and fresh. There are a number of potato varieties available. 'Desiree' and 'King Edward' are popular main crop choices. Try a late variety, such as 'Maris Peer,' which needs to be planted in late summer and should be ready to harvest and eat around midwinter.

Before you plant, lay the seed potatoes in single layers on a tray, in egg cartons, or in a shallow box and leave in a dark place to "chit" (sprout shoots). Once your seeds have shoots, they are ready for planting. Don't put them out too early unless you intend to grow early potatoes, in which case you should cover them with a cloche or plastic covering of some kind.

During the war years, in order to save money, people planted cut-up pieces of potato, which often grew into healthy plants, producing many new organic potatoes. However, this method is a little hit and miss because today we tend to buy plastic-bagged potatoes that have had a certain amount of processing along the way.

Transplanting

If you have space in your vegetable patch, it's definitely worth growing a couple of rows of potatoes. Don't choose an area where tomatoes, peppers, or potatoes have been grown in the previous couple of years because these vegetables all belong to the same family and can spread diseases to each other through the soil.

You'll need a good depth of soil, and you should dig over the ground well before planting. Remove any perennial weeds, large stones, and nonorganic debris from the soil and then dig a trench. You can find manual or mechanical tools for making trenches, but they aren't absolutely necessary.

Use whatever feels comfortable for the job; sometimes, a simple spade will do the job well, but if your soil is a little crumbly, it may be better to find another way. Dig your

Use your potatoes in hearty dishes throughout the cold-weather months.

trench about 6 to 8 inches (15 to 20 cm) deep, and if you are planting more than one row of potatoes, leave about 3½ feet (1 m) between rows.

I started one of my best potato crops ever on a bed of comfrey. If you have comfrey available, pick the leaves and lay them in the bottom of the trench. Sprinkle a little soil over the top and then place your seed potatoes, shoots up where possible, on top of that. Allow about 12 inches (30 cm) between each potato. The comfrey leaves help release nutrients from the soil into the roots of your potato plants. If you do not have any comfrey leaves, it's not the end of

the world. Simply lay your seed potatoes along the bottom of each trench, again 12 inches (30 cm) apart. Cover with the soil you dug out and water the ground well.

Keep potatoes weed free, and water them regularly if the weather is dry. After a few weeks, you will notice small, bushy leaves emerging along the line. Try to wait until all the plants are up and showing if you can (there may be a "dud," so don't leave it for too long).

At this point, you should "earth up." Using a heavy rake or any other tool that works for you, pull the soil from both sides of the line and cover the newly emerging plants completely. Leave the soil loose on top of the mound and the plants. You will now have shallow trenches on either side of a long mound.

Make sure the ground doesn't dry out, and remove the weeds as they emerge. A few weeks later, once the potato plants have poked through again, repeat the whole earthing–up process, and then repeat it again a few weeks after that.

Once the plants have been earthed up three times in total, you will have deep trenches on either side of a fairly large mound. Let the plants grow and develop from this point. The earthing–up process gives the roots a chance to expand, and you will get far more potatoes using this method. You can also grow potatoes in a raised-bed system, although they tend not to crop as well if you don't earth them up.

Stored in a cardboard egg carton, these half-dozen seed potatoes have produced chits, or shoots.

Care and Maintenance

If the weather is particularly warm and muggy, your potato plants may be susceptible to the blight virus. Many growers spray their potato crops every few weeks during the growing season to prevent blight attacks. Use a Bordeaux mixture (copper sulfate and hydrated lime dissolved in water) or other organic product.

When the plants have flowered or are just starting to flower, treat yourself to a few tiny new potatoes. Scrape the earth away from the side of the mound next to a healthy plant and search for a few baby potatoes. Push the earth back once you've taken one or two potatoes from a plant. Don't take more than one or two potatoes from each plant. The potatoes left behind will keep growing and developing as long as you cover them again properly with the soil.

Harvesting

When your potato plants have died back completely, the potatoes underground will have more or less finished growing. In the morning on a dry day, carefully fork around the plants, allowing a good area all around them because the potatoes spread out; if you are very careful, you may not spear too many. It's almost impossible not to catch one or two with your fork, but don't try to store these; simply use them first. Cut away the rough-looking parts and have the rest for dinner.

Once you have dug around a plant, gently lift the soil, and you'll be able to find the potatoes. The best way to do this is with your hands. Wear gloves to protect your skin if you need to. Push away all of the earth and make sure that you remove all of the potatoes from the ground. Repeat the digging and lifting process with each potato plant and then spread the potatoes over the dry earth for a couple of hours in the sun, turning them as they dry. Don't leave them out overnight. When the skins are dry, you can store your potatoes.

Storing

Years ago (and, in some cases, still today), root vegetables were kept in specially prepared earth clamps for the winter. There's no reason why you shouldn't try using this method of storage, especially if the upcoming winter isn't expected to be too wet.

Smooth out an area of ground where you just dug out your potatoes or other root vegetables and firm the soil with your feet. Dig a shallow trench around a circular area to deter hungry animals. Lay an 8-inch (20-cm) depth of dry straw on the firmed-down circular area and pile your vegetables in the middle of the straw to form a cone or pyramid shape, first making sure that the vegetables are completely dry.

Cover your vegetables with straw and then cover the whole pile with earth, leaving a 4- to 6-inch (10- to 15-cm) ventilation hole at the top. Stuff this hole with straw. In a properly built clamp, your vegetables will keep for a few months.

You can successfully store potatoes for many months as long as you keep them out of direct light and they are not too damp or cold. Potatoes, like many vegetables, have a high water content and will rot if allowed to become wet or frozen. Either build a vegetable clamp as previously instructed or store your potatoes in single layers in stackable boxes. Always make sure that the skins are thoroughly dry before placing the potatoes in the box, and never store in direct light or damp conditions. You can buy potato storage units in garden centers or through other garden-equipment suppliers.

Nutrients

Potatoes are a valuable source of vitamin C and some B vitamins. They are also a good source of carbohydrates. In the past, potatoes have been thought to aid digestive problems as well as dry skin and sores. They have only a trace of fat in their natural state. Potatoes with their skins are considered to be more nutritious than potatoes without skins, but no matter how you prepare it, the honest spud makes a valuable contribution to the family diet.

Small potato plants emerging from the soil are ready to be earthed up (completely covered with soil).

RECIPE IDEAS

Potatoes are probably one of the most versatile vegetables you can grow. You can make many dishes and freeze them for the winter months ahead, when the potatoes in your storage boxes or clamps are starting to get a bit too soft to use.

Fresh potatoes straight from your garden taste like no other potatoes. Steamed or lightly boiled and served with a sprig of mint, they will complement any roast. When you first dig them up, the skins will probably just rub off. Try not to peel your homegrown potatoes (unless, of course, you did not grow them organically).

BAKED POTATOES

One of the most popular ways to serve medium to large potatoes is baked. Simply wash the potatoes before baking them in an oven on medium heat or until they are cooked through. Serve with one of the following or with any favorite filling:

• Plenty of butter, salt, and black pepper

• Grated cheddar and coleslaw

• Baked beans

• Tuna and corn with mayonnaise

Alternatively, split them, scoop out the flesh, and mash with cheese and butter before spooning back into the skins. Grill for five minutes to brown. Be sure to scoop out the flesh very carefully so you don't damage the skins too much.

SCALLOPED POTATOES

Creamy scalloped potatoes are particularly delicious served with filet mignon and a crisp green salad, and this is also a fantastic way of serving potatoes as accompaniments to a variety of dishes.

• Preheat your oven to 375 degrees Fahrenheit (190 degrees Celsius). Grease a large, shallow baking dish with butter.

• Slice 2¼ pounds (about 1 kg) of potatoes as thinly as you can, using a vegetable peeler or a very sharp knife. Layer them evenly in the buttered baking dish, along with a finely sliced onion, seasoning as you go.

• Pour 10 fluid ounces (284 ml) of heavy cream and 2½ fluid ounces (75 ml) of milk over the potatoes. Dot the potatoes with butter and cover the dish with foil.

• Place the baking dish on a baking sheet and bake for 1 hour. Discard the foil and bake for a further 15 to 20 minutes or until the potatoes are fully cooked and golden brown.

POTATO AND CHICKPEA CURRY

Potatoes work well in curries, and this easy recipe takes just 25 minutes to whip together. The dish is packed with protein, so it is a nutritious choice for vegetarians. If you prefer a less spicy curry, simply add less curry paste and go easy on the cilantro.

• Cook two roughly chopped, medium-sized potatoes in salted boiling water for 10 to 15 minutes until tender.

• Heat a little oil in a frying pan, add a finely sliced onion, and cook for 5 minutes until soft.

• Stir in 1 to 2 tablespoons curry paste, 10 ounces (300 g) drained chickpeas, and 8 fluid ounces (250 ml) cold water. Simmer for 5 minutes.

• Drain the potatoes and add them to the frying pan along with a small bunch of chopped fresh cilantro. Stir over low heat for a few minutes.

• Serve with steamed rice or naan bread.

OTHER IDEAS

• Spread mashed potatoes over shepherd's pie, sprinkle with cheese, and brown under the broiler.

• Form cooled mashed potatoes into little balls to make croquettes. Coat in flour and fry in hot oil or bake in the oven for 10 to 20 minutes.

• For a treat, make homemade french fries and potato chips. For fries, cut washed potatoes into chunky pieces and fry in hot oil until golden brown. For chips, use a potato peeler or mandoline to slice potatoes very thinly and then fry in hot oil for a couple of minutes.

• Boil, steam, or roast potatoes or add to stews, soups, and quiches for extra bulk and flavor.

Carrots

Carrots have been cultivated for thousands of years, and people grew many different colors of carrot in their gardens—even purple carrots had a place on the dinner table. In the sixteenth century, Dutch growers crossed a yellow and a red variety and produced the bright orange carrot we know and love today. It is possible to buy special seeds for heirloom varieties, although the everyday seeds seem to do the job perfectly well if you want a crop of orange carrots. If you have the space, time, and inclination, though, try a different-colored variety as well.

Seed

It's always a good idea to buy your seeds from a reputable supplier. Cheap seeds can be old or damaged, and it will be a struggle to get any decent crops from them. Carrots are a root vegetable and require a certain soil depth to fully develop. However, there are shorter, fatter varieties available that will grow well in a shallow depth of decent topsoil. Read through the manufacturer's growing recommendations on the seed packet, preferably before you buy seeds but definitely before you sow.

If you like the idea of growing different-colored carrots, you may have to put forth a little more effort to find a suitable variety, but it's worth it. Again, check the depth of soil required for the carrots to fully develop.

Always sow your carrot seed in situ. Carrots, like most root vegetables, don't transplant well. It is possible to transplant them, but the carrots will very likely "fork," or split, and twist into unusual shapes.

Sowing

If your ground is tired or lacking in nutrients after being overworked in recent years, dig in some organic compost or well-rotted manure a month or two before a spring sowing. Don't add fresh manure to your soil just before or after sowing carrot seed because it will, like transplanting, cause the carrots to fork or split.

When you are happy with the soil, dig it over as soon as it is workable in the spring. Remove any perennial weeds, large stones, and nonorganic debris. Large stones or lumps of concrete in the ground will inhibit the growth of your root crops. Fork after digging to break up the soil and then rake it over to a fine consistency.

As soon as the weather warms up, you can sow your carrot seed, but find out the recommended sowing times for your particular variety because some carrot varieties should be sown earlier than others.

Sow very short lines of seed, up to 3½ feet (1 m) long, and rake the soil carefully over the top. Water gently and keep watering from time to time if there isn't at least a little daily rain. The seeds won't germinate if they dry out. They also won't germinate in waterlogged soil, so make sure that the soil is well drained, and don't overwater.

It's best to sow very short lines of carrot seed because once the seedlings are about an inch (a couple of centimeters) high, you will need to thin them out. This is a

Don't be afraid to get colorful with your carrots!

laborious job if you have sown long lines of seed. If you can, thin out your seedlings when it is raining to deter the carrot fly that can damage your crops; carrot scent is less pungent to the carrot fly when it is raining. Otherwise, water the ground first and then carefully remove weaker seedlings, allowing about an inch (a couple of centimeters) between the remaining plants. Firm the soil gently around the remaining plants with your fingers. Water if necessary and allow them to keep growing for a few more weeks and then repeat the process. During the second thinning session, you should be pulling up very tiny carrots. These are edible and delicious, so throw them in your salad bowl rather than on the compost heap.

Care and Maintenance

Thin out your carrots at least once or twice per line to allow space for the roots to develop. Sow short lines of seed every couple of weeks so you have a constant supply, but not so many that the maintenance becomes overwhelming. It may seem like a lot of work to produce a few carrots, but the process isn't hard, and it gives you a good excuse to spend time in the garden for ten minutes every now and then. And, of course, homegrown carrots are superb!

The seeds can take some time to germinate; anywhere up to six weeks is possible. Make sure that you water your plants and keep them free of weeds during the growing season. Carrots don't like to dry out; otherwise, they are fairly hardy

Carrot varieties are not just orange; they come in a wide variety of colors, shapes, and sizes.

plants. A little gentle surface hoeing every now and then will help aerate the soil and remove any weeds.

Don't feel as if you have to wait until each plant has fully developed before you pull them to eat. Always try to lift carrots from the row when the soil is damp to avoid tugging at the others too much. Loosen the soil gently around the plants with a fork if necessary.

Harvesting

Pull your carrots as you need them right through the growing season. You should harvest them all before the heavy autumn rains, and definitely before a ground frost, unless you have a particularly hardy variety.

Check your seed packet for growing, harvesting, and storage advice.

Loosen the soil around your carrots before pulling them from the ground. If they feel too firm to pull up, dig a little deeper before trying again. If you force the root, it may break. Let the carrots dry in the sun for an hour or two before bringing them in to store.

Storing

You can store carrots in a vegetable clamp or spread them out in trays and keep them in a cool, dark place for several months. Hanging them in sacks can also work well. You can also store carrots in barrels of sand, but that isn't always a viable option in a modern household these days. If you do try the barrel method, watch out for mice.

Never wash carrots before storing; they will rot very quickly. Simply dry them in the sun or a dry place and then brush off any excess soil before storing.

You can freeze carrots, although they will lose some of their texture and taste. Freeze them sliced as soon as possible after harvesting, and remember to label them before storing. Canning is another option if you have the equipment to do so properly. Always store canned vegetables out of direct light.

Nutrients

During World War II, pilots were given carrots in the belief that it would make them better able to spot their targets; it became a well-known saying that you should eat your carrots if you want to see in the dark. Although it sounds like an old wives' tale, there is some evidence to support the fact

GARDENER'S TIP

Plant carrots and onions in lines next to each other. You will find that the scent from the onions deters the carrot fly, and the scent from the carrots deters the onion fly.

that carrots actually do promote good eyesight. Carrots contain vitamins A, B1, and B6 and significant quantities of calcium and iron. Vitamin A is essential for good vision, and one medium-sized carrot can provide all of your recommended daily intake of this vital vitamin.

These rows of young carrot plants have been thinned already and may need thinning again.

RECIPE IDEAS

The carrot is one of the most versatile vegetables. Carrots can be eaten in sweet and savory dishes, either cooked or raw. Carrot sticks make a healthy, satisfying snack. You can grate carrots or peel them into thin ribbons as a garnish for a salad or a hot dish, or you can serve them on their own as a tasty side dish. And carrot cake is one of the most popular sweet treats.

QUICK AND EASY COLESLAW

Mix grated carrots with grated white cabbage and finely chopped or grated onion and then stir in a mixture of equal parts mayonnaise and plain yogurt. You might try adding a few chopped walnuts, pine nuts, or sunflower seeds for added crunch. Chill for 20 to 30 minutes before serving.

GARLIC CARROTS

• Scrub carrots well, rinse under running water, and peel (although, if you prefer, you do not have to peel carrots that you've grown organically).

• Slice into coins or matchsticks and steam or boil for about 10 minutes or until tender. Remove from the heat and drain well.

• Return carrots to the pot and stir in 1 crushed garlic clove, 1 tablespoon chopped parsley, and 1 to 2 tablespoons butter.

• Cover the pot, shake it to distribute the ingredients, and let the carrots sit for a few minutes to absorb the garlic flavor. Serve hot.

HERBED CARROTS

Prepare in the same way as garlic carrots but instead of adding garlic, add 1 to 2 tablespoons of your favorite chopped fresh herbs, depending on what you are serving the carrots with. For example, use chopped fennel or dill leaves to complement a fish dish. Add mint if you are serving the carrots with roast lamb. Cilantro also works well with carrots and will add color and flavor to almost any meal.

CARROT AND CILANTRO SOUP

Soups are great for feeding the family during the cold winter months, and carrot soup is no exception. There are many recipes for carrot soup, but this one is simple to make and will appeal to most people's tastes. Use the cilantro sparingly unless you like your food on the spicy side.

• In a large pan, warm a little oil or butter and then add 2 finely chopped onions and 6 cleaned and sliced carrots. Stir and cook gently for 5 to 10 minutes or until the vegetables are soft. Stir frequently to prevent them from burning.

• Add about 2 pints (1 liter) of vegetable stock, season with salt and pepper, and cover the pan with a lid. Bring to a boil and then reduce the heat and simmer for about 20 to 30 minutes.

• Stir in a handful of chopped fresh cilantro and bring back to a boil. Reduce the heat again and simmer for another 10 to 20 minutes until thoroughly cooked. Check the seasoning and add a little more cilantro or pepper if required.

• Remove from the heat, cool slightly, and then blend in a food processor or blender until smooth. Return to the pan and heat through very gently, stirring constantly. Serve hot in individual bowls with a swirl of light cream or crème fraîche on top and garnish with a sprig of fresh cilantro.

• If you don't have a blender or food processor, drain the vegetables, retaining the liquid, and then mash the vegetables with a potato masher. Stir the mashed carrots together with the reserved liquid to form a smooth, creamy soup and then reheat as previously mentioned.

• You can freeze the soup (minus the cream) in containers; it will keep for several months in the freezer.

OTHER IDEAS

• Add carrots to stews and casseroles to provide bulk, flavor, and plenty of nutrition.

• Cut carrots into strips and stir fry with green peppers and bean sprouts.

• Finely dice carrots and add to bolognese sauce and lasagna.

Rutabaga

Rutabaga is a delicious vegetable that is sometimes overlooked. It is hardy and will stay in the ground right through the winter with a little care. A close relative of the turnip, rutabaga is classified as a root vegetable and is also a member of the *Brassica* genus. It tends to grow larger than the turnip and usually has a purple-colored area at the base. There are a few different varieties of rutabaga available.

Known as *swede* in some parts of the world, such as the United Kingdom, the rutabaga doesn't have as long a history as other root vegetables, but, by the eighteenth century, it had become an important garden crop in many parts of Europe.

Rutabaga will benefit from well-nourished soil. Dig in plenty of organic well-rotted compost or manure during the month or two before you sow the seed. Never put fresh manure or compost on the area before or after sowing because this will cause root crops to split, and they won't develop well. You can also spread well-aged compost or well-rotted manure over the ground in the autumn before planting. Rutabaga is generally sown in late spring and early summer, so the ground should have plenty of time to settle.

Seed

There are a few different varieties of rutabaga seed available from good garden suppliers. If possible, buy a variety resistant to clubroot, a disease that can affect all brassica plants and will destroy the roots.

Because rutabaga is a brassica, it doesn't like acidic soil, so the pH balance should be about neutral. Further, acidic soil will contribute to the possibility of clubroot, so make sure that your soil is well balanced before you start growing this root vegetable.

As with all root crops, you should sow seed in situ; rutabaga also can be successfully grown using biodegradable modules in which you transplant the entire pot to avoid any contact with or damage to the roots.

Sow seed from late spring onward. Check the growing recommendations on your seed packet because varieties and regional needs may differ.

Sowing

Dig over your plot and remove any large stones, perennial weeds, and nonorganic debris, and then rake over to a fine consistency. The area should be well drained and in a fairly sunny spot. If you have trouble getting your ground well drained, form a ridge with the soil and sow your seed on top of the ridge. The water will then drain away from the plants, which reduces the possibility of waterlogging.

Sow seed thinly along a line and leave about 24 inches (60 cm) between lines. Remember that you will need to allow around 10 inches (25 cm) per plant eventually, so be sure to prepare fairly long lines if possible and sow seed very thinly. When you choose the spot in your garden for rutabaga, consider that they can stay in the ground for much of the winter and plan accordingly.

Cold-resistant rutabagas can stay in the ground until you need them.

Prepare a ridge with the back of a rake along a line to sow rutabaga seeds. Sow thinly and thin out the seedlings.

Rutabaga seed tends to germinate within a couple of weeks. When the seedlings are about an inch (a few centimeters) high, you will need to thin them. Make sure that the soil is damp before thinning the plants to avoid pulling out plants that you want to leave in the ground. Work along the row, removing weaker plants and leaving 10 inches (25 cm) between each plant to give them space to develop (unless otherwise instructed by recommendations on the seed packet). Don't attempt to replant the seedlings that you pull out. Although it seems like a waste to remove them, they won't grow well if you replant them.

Care and Maintenance

When your plants are small, you should protect them from slugs, snails, and flea beetles. Flea beetles arrive early in the year, so if you can hold off on sowing your seeds until late spring, you may miss them. Slugs and snails, however, are always around, so do what you can to stop them from attacking your young crops.

The other problem you may have with wildlife is the cabbage white butterfly. It lays its eggs on the underside of the leaves of brassica plants and, within days, the caterpillars hatch and devour everything in their path—usually your valuable crops. Keep an eye open for the beautiful but very crop-damaging cabbage white butterfly. Check on the underside of your plants' leaves every day for eggs. If you find them, either scrape them off or remove the entire leaf and then compost it.

Keep the ground watered during dry periods and remove any weeds as they appear. If you are using biodegradable pots, sow a few seeds in each pot and keep the soil damp but not waterlogged. When the seedlings are about an inch (a few centimeters) high, carefully remove the weaker ones, leaving one strong plant per pot, and then transplant the whole pot, leaving 10 inches (25 cm) between each plant to grow.

Harvesting

Rutabaga can be stored for a couple of weeks after pulling, but they are so hardy and tolerant of frost and cold that you can leave them in the ground right through the winter (however, research your variety just in case you have a frost-tender type). Fork around the whole plant carefully before attempting to pull up the root. You can use rutabagas as soon as they are about the size of a grapefruit; however, after the first frost, they will be sweeter and more enjoyable. Take advantage of their resistance to the cold and leave your rutabagas in the ground until you need them.

Storing

Although this is one of the vegetables that you can leave in the garden through the winter months, or until the rain gets too heavy in early spring, rutabaga will keep for several weeks after harvesting if required.

Dig them up and allow them to dry outside for an hour or two. Rub off any excess soil and store in a dry, dark place until you need them. Try to use them within a week or two. Some rutabaga varieties may store better than others, so check the storage recommendations on your seed packet. Otherwise, if you have to dig them up, keep them for only a week or two before using. You can also store them in the vegetable compartment of your refrigerator for a couple of weeks.

Nutrients

As with all root vegetables, rutabaga is rich in minerals, including calcium, magnesium, manganese, and potassium; it also has very low fat content, boasts vitamins A and C in its composition, and is a good source of

In the foreground, a rutabaga crop is growing in a raised bed. Note the protective netting in the background.

dietary fiber. Root vegetables are nourishing in the winter months, and you should eat them regularly.

Rutabaga has a high natural sugar content, especially after the first frost of the year. Enjoy your first rutabaga after the first frost in a hearty stew or sustaining casserole.

RECIPE IDEAS

Traditionally, rutabaga has always been boiled and mashed, and it can be served with almost any main course. Its sweet flavor will complement many a meal, including your favorite Sunday roast. Rutabaga can also be eaten raw; one idea is to grate it very finely and mix it into a winter salad for extra vitamins and crunch.

HERBED MASHED RUTABAGA

• Peel and cut your rutabaga into chunks.

• Put into a pan, cover with water, and bring to a boil. Reduce the heat and simmer for 20 minutes or until the rutabaga is tender and fully cooked. (You can steam instead of boil if you prefer).

• Once cooked, drain the rutabaga well and return it to the pan. Mash using a potato masher or a fork, stir in a tablespoon of butter, and add black pepper to taste and a handful of finely chopped fresh herbs of your choice.

• Mash again to mix well and serve hot.

RUTABAGA AND BACON

Rutabaga and bacon were made for each other, and this is a great way to use up any uneaten rutabaga.

• Peel and cut your uncooked rutabaga into small cubes. Cook about 6 slices of bacon (trimmed of fat) in a little oil and then remove the bacon. Drain on paper towels to remove excess oil.

• Peel and finely chop an onion and add it to the oil in the pan. Cook over a gentle heat until the onion is soft and translucent.

• Add the rutabaga cubes and stir well. Cook over low heat, stirring constantly until the rutabaga is tender.

• Cut the bacon into pieces and add it to the pan with a handful of chopped fresh parsley. Continue stirring and cooking gently until hot. Serve with mashed potatoes.

• Note: You can use leftover ham instead of bacon. When reheating meat, make sure that it's piping hot all the way through before serving.

CORNISH PASTIES

This traditional British delight was originally created for Cornish tin miners who, unable to return to the surface at lunchtime, would have this hearty pastry stuffed with meat and rutabaga. This classic is just as popular today in the United Kingdom in lunchboxes, for picnics, and as main dishes accompanied by green vegetables. There are endless variations for filling, but the traditional version uses beef, potato, and, of course, your delicious homegrown rutabaga. The following makes four pasties.

• Preheat the oven to 425 degrees Fahrenheit (220 degrees Celsius).

• In a bowl, mix together the filling ingredients: 12 ounces (350 g) rump or braising steak, finely sliced; 1 large onion, finely chopped; 2 medium potatoes, finely sliced; and 1 medium rutabaga, finely diced. Season with salt and pepper.

• Divide 13 ounces (375 g) ready-made pastry crust into four rounds and roll each out to about 9 inches (23 cm) across, using a dinner plate to trim to shape. Firmly pack a quarter of the filling along the center of each round, leaving a margin at each end. Brush the pastry around the edge with beaten egg, carefully lift up both sides to meet at the top, and then pinch the edges together to seal.

• Carefully lift each onto a baking tray, brush with the remaining beaten egg, and put in the preheated oven. After 10 minutes in the oven, turn the temperature down to 350 degrees Fahrenheit (180 degrees Celsius) and bake for another 45 minutes or until golden.

"NEEPS AND TATTIES"

Rutabagas are called "neeps" in Scotland, and "tatties" are potatoes. This rutabaga and potato dish is the traditional accompaniment to haggis. It is best to make it the day before you want to serve it.

• Preheat the oven to 400 degrees Fahrenheit (200 degrees Celsius). Cut 4 large potatoes into chunks and cook in a pot of boiling salted water for 5 minutes. Drain the potatoes, return them to the pan, and heat them again for up to a minute to dry them out.

• Pour 3 tablespoons of oil into a large roasting pan and heat in the oven until hot but not smoking. Stir the potatoes into the hot oil and cook in the oven for 50 minutes, turning occasionally.

• Meanwhile, roughly chop a medium rutabaga and cook in boiling salted water for 50 minutes or until soft and tender.

• Drain the rutabaga pieces and add them to the roasted potatoes. Mash together, keeping everything chunky, and allow to cool.

• Before serving, dot with a little butter and return to the oven for 25 to 30 minutes.

OTHER IDEAS

• Rutabaga is a perfect vegetable to add to stews and casseroles. Add it early in the cooking time because it generally takes longer to cook than carrots or potatoes.

• The leaves of the rutabaga plant can be eaten as well as the root. Don't strip a plant completely of its leaves, or else the root won't develop, but you can steal a few leaves here and there from time to time. Steam or boil as you would spinach and serve hot as a side dish.

Turnips

A traditional winter crop, turnips have been cultivated for thousands of years and were even used as currency in trading. This root vegetable was very common before the potato became more popular, pushing the turnip to second place. Turnips are an excellent crop to grow later in the year, but it won't grow well in long, hot summers; in fact, they tolerate the first cold nights of autumn, and even winter, very well.

Turnips are best eaten when fairly small although they are good even when they have reached a more substantial size. Like rutabagas, turnips are wonderful in hearty winter dishes.

Seed

Turnips can be grown for their greens as well as the roots, but care should be taken not to pick too many leaves off each plant. Different varieties can be grown specifically for the root or for the greens. Choose your seeds carefully, according to your own needs.

As with other root crops, turnip seed should be sown in situ in well-prepared ground. It is possible to sow turnips in individual pots as long as they are biodegradable. You can transplant the whole pot later to avoid damaging the roots or having to remove the seedlings from the soil.

If you are using this method, when the seedlings are about an inch (a few centimeters) high, wet the soil in the biodegradable pots and then very carefully remove the weaker seedlings. Leave behind one strong plant per pot to grow on. It

is, however, preferable to sow your turnip seed directly into the ground where you plan to grow your crop. Check your seed packet for specific instructions pertaining to your variety.

Sowing

Prepare your ground well. Choose a sunny, well-drained spot in the garden and dig it over to a good depth to give your root crops a chance to develop properly. Remove any perennial weeds, large stones, and nonorganic debris. Don't dig in any manure or compost unless it is well rotted, and ideally you should incorporate this into the soil at least a month or two before you plan to sow the seeds.

Rake over the soil to a fine consistency, make a shallow drill, and sow your seeds thinly along the row. Leave about 12 inches (30 cm) between rows. The seed can be sown in early spring for summer varieties or later on in the summer for autumn and winter types. Turnips seem to taste better later in the year and, because they are the kind of vegetable we tend to eat in the winter months, it's preferable to sow seed in the summer months rather than in the spring. Always check your seed packet for variations.

When your seedlings are about an inch (a few centimeters) high, you should thin them out so that there are about 4 to 6 inches (10 to 15 cm) between each one to allow root development. Again, check the

Both the leaves and the roots of turnip plants are edible.

Turnips are brassicas and are vulnerable to the same diseases as cabbages and other brassicas. Avoid clubroot by regulating acidic soil, as mentioned in the Gardener's Tip on this page, and also watch out for the cabbage white butterfly, which will lay eggs on the underside of your brassica leaves. Turnips are not immune to this butterfly, and the resulting caterpillars can completely annihilate your crops in just a couple of days. Be vigilant. Check the undersides of the leaves every day as soon as you see the first cabbage white butterfly. Remove any eggs by scraping them off or removing and composting the whole leaf.

seed packet's growing recommendations for any variations. Make sure that the ground is wet before you start and remove the weaker plants where possible. Firm down the remaining plants with your fingers very carefully and then water gently to allow any movement to settle. You can add the plants that you pull out to your salad bowl. They are very nutritious, and it's a shame to discard them. Don't be tempted to transplant these seedlings; as with most root vegetables, they will fork or split and won't develop well.

Care and Maintenance

Keep the ground watered, especially during long, dry periods, and ensure that the area is free from weeds. If you let your plants dry out, they won't thrive, so water them regularly.

Harvesting

Pull up and enjoy a few turnips when they are still relatively small. Many vegetables are sweeter when young, and you can freely enjoy them when you are growing your own. Don't pull them all when they are small, however—allow some to develop fully.

Turnips will be dirty when pulled from the garden; be sure to brush off the soil before storing them.

You can use turnip leaves like you would use spinach or rutabaga leaves. Take only a few leaves from each plant or grow a variety specifically bred to produce plenty of foliage but not fully grown roots.

Storing

You can store turnips successfully in a vegetable clamp as described at the beginning of this chapter in Potatoes (see page 43). They will keep for many months in ideal conditions.

Laying turnips in boxes, once you've cleaned the soil from the turnips, is another storage option. Don't wash turnips; like other root vegetables, they will not keep if you do. Just dry the roots in the sun or a cool airy place and then brush off any excess soil. They should keep well for a few weeks without wrinkling or drying out too much.

You can also freeze turnips; cube and blanch them in boiling water for five minutes and then drain, cool, and freeze them quickly

Perfect turnips, ready to be used immediately or stored for later use; they keep very well for several months.

in a labeled container or freezer bag. Or make and freeze a batch of turnip soup.

Nutrients

Turnips are a good source of vitamin C, with around 11 or 12 mg per 4 ounces (100 g). They are also very low in calories, making turnips an ideal vegetable in calorie-controlled diets. A cooked turnip has only about 22 calories per 4 ounces (100 g).

As with most root vegetables, turnips are a good source of minerals, including calcium and potassium, as well as dietary fiber, so include this useful and tasty vegetable as often as possible in your winter diet. Turnip greens are also rich in dietary fiber, minerals, and vitamin C, and they contain a significant amount of folate. However, turnip greens contain goitrogens and shouldn't be eaten by those with untreated thyroid conditions.

RECIPE IDEAS

People often think of the turnip as a boring, rather bland-tasting vegetable, but with a little inspiration in the kitchen, turnips can be very tasty. This root is definitely worth eating for its nutritional value alone.

MASHED TURNIPS AND TURNIP CROQUETTES

You can mash turnips in the same way as you would potatoes and rutabagas. Prepare them by peeling and cutting into fairly small pieces. Potatoes generally cook a bit quicker than turnips, so if you want to cook them together in the same pot, cut the potatoes into larger pieces than the turnips.

• Put all prepared potatoes and turnips in a pot and cover with water. Bring to a boil and then reduce heat and simmer until all vegetables are fully cooked.

• Remove from the heat and drain well. Return to the pot and mash with a potato masher or a fork. Stir in a tablespoon or two of butter, a little seasoning, and a handful of chopped fresh herbs—sage and thyme work well with turnips.

• Stir well and serve immediately.

• You can store any leftovers in the fridge overnight and make them into turnip croquettes the next day. Shape the mixture into balls and flatten them to about 1 inch (2.5 cm) thick. Fry gently in a little oil on both sides for 5 to 10 minutes or until browned and hot all the way through, or bake them on a baking sheet in an oven at medium heat for about 20 minutes. Serve hot with a salad.

CRUNCHY TURNIP CRUMBLE

• Boil, drain, and mash 2 medium turnips.

• Mix together 1 tablespoon brown sugar, 1 teaspoon baking powder, salt, pepper, and a pinch of nutmeg. Add the mixture to the mashed turnips and then place in an oven-safe dish.

- Toss 8 tablespoons breadcrumbs with 2 tablespoons melted butter and sprinkle on top of the turnip mixture.

- Bake in the oven for 20 to 25 minutes at 350 degrees Fahrenheit (180 degrees Celsius).

CREAMY TURNIP GRATIN

- Rub two small ramekins with garlic and butter generously.

- Peel and thinly slice a medium turnip, parboil for 2 minutes, and drain. Return it to the pan and add 5 fluid ounces (150 ml) heavy cream, 1 teaspoon dried herbs, and 2 ounces (50 g) thinly sliced pancetta.

- Season and simmer until the turnip is tender.

- Spoon the mixture into the prepared ramekin dishes and warm in the oven until golden on top.

OTHER IDEAS

- Finely grate a raw turnip and add to a salad.

- Roast as you would potatoes or cut into very thin matchsticks and stir-fry.

- Grate into the sauce when you are making lasagna, bolognese sauce, or a casserole.

- Make into a soup with carrots, potatoes, and leeks. Add a splash of cream before serving.

Beets

Beets have been cultivated for many centuries, although before the seventeenth century, they were grown mainly for their leaves in home gardens. The roots were used for medicinal preparations; among other uses, beets were considered an effective aphrodisiac because they contain the element boron, which has proven to have an effect on hormones.

The round purple-red roots we eat today are just as nutritious as the leaves. If you like the idea of using beet leaves (called *beet greens*) in place of spinach or other leafy greens, grow a line of beets just for the foliage and another for the roots.

It appears that the beet was originally a seaside plant found over large coastal areas in Europe and Asia. The plant is well adapted to the vegetable garden and is probably one of the easier root crops to grow.

Seed

When you buy beet seeds, make sure that you know what you're buying! Some beet seeds will be better for foliage than for roots. Try both if you have the space or, if you are interested in only the root crops, select a good-cropping everyday beet to start with.

If you can't find a variety specifically for the foliage, grow a small line of any beet seed and use some of the plants for leaves and allow the rest to develop roots. When you remove leaves regularly, the plants will produce more leaves and won't put energy into developing the roots. The seeds are "multiple" and will need to be thinned out later, no matter how thinly you sow.

It may be possible to sow the seeds in individual biodegradable pots, but it's generally better for all root crops to be sown directly outside in the vegetable patch. If you transplant root-crop seedlings, they are likely to fork or split and won't develop into healthy roots. If you do try using biodegradable pots, remove the weaker seedlings to leave a strong seedling in each pot as soon as they are large enough to handle. Soak the soil with water first so you don't dislodge the plants that you are leaving in the pots. Firm the soil gently around the plants after thinning and then water gently.

Sowing

When sowing outdoors, prepare your soil first. Beets don't like acidic soil, so check the pH levels of the soil if you aren't sure. If necessary, add lime or another organic material to rebalance the soil a month or so before sowing the seed.

Choose a sunny, well-drained spot in the garden and dig over to a good depth. The cleaner and more prepared your soil, the better your root crops will develop. Remove any perennial weeds, large stones, and nonorganic debris and then rake over the surface to a fine consistency for your seeds.

Sow beet seeds after all danger of frost has passed. With a raised bed system and a plastic cloche-type covering, you could start beet seeds a little earlier, but the seeds won't germinate well if it's cold or if they

With their purple-red roots, beets are easy to recognize.

Harvest your beets before they get too large. Loosen the soil around the roots first and then pull them out by their stems.

Keep beet plants healthy by watering regularly in dry weather and making sure they are weed free. Beets are a hardy root crop and fairly resistant to bugs and disease, although all young plants will need protection against slug and snail attacks.

Harvesting

You can pull and eat the roots as soon as they are the size of a golf ball. Make sure that you harvest all of them before the first frost in the autumn or winter unless you have a frost-resistant variety. Try not to leave them too long because they can become woody and lose some of their flavor. Pull them when they are small if you intend to pickle them.

Use a fork to gently loosen the soil around each beet before you lift it to prevent any damage. Remove the leaves by twisting them rather than by cutting through the root. You should collect leaves regularly, but never strip any one plant of all its leaves. Take a few leaves from each plant, and the plants will stay healthy and crop for longer in the season.

Storing

Beets are often pickled for storage, or you might want to can them as you would can any other vegetable. Canning requires special equipment, but you can pickle beets with just a large pan and some jars. See the

don't have enough light. Check the growing recommendations on your seed packet before you start so that you're aware of any variations for your particular region or variety of beet.

Sow seeds thinly in shallow trenches, preferably positioning one seed cluster about every inch (few centimeters) or so, leaving about 12 inches (30 cm) between rows. Cover with soil. Water gently after sowing and keep the soil watered regularly in dry weather. Remove any weeds as they appear.

Care and Maintenance

After a few weeks, your seedlings will need thinning out. Choose a wet day or soak the ground first with water. Gently remove weaker seedlings, leaving 4 to 6 inches (10 to 15 cm) of growing room for each plant. Again, double-check your seed packet for spacing advice. Firm the soil gently around the plants left in the ground. Water again if needed.

Condition the soil well before sowing to ensure a good crop of beets and then sow the seeds in shallow trenches.

recipe ideas on page 69 for a quick and easy way to pickle beets.

You can store the roots in the refrigerator for about a week or in a cool, dark place for a few weeks. Keep an eye on them because they don't tend to store as well as other root vegetables. You can also store beets in a vegetable clamp (see page 43) if you are using this method to store root crops.

Nutrients

Beets are high in dietary fiber, folate, and vitamin C, and the dark pigment actively helps fight free radicals in the body. Beets have been consumed in soup form for many years and are sometimes believed to be the secret ingredient to longevity. Packed with minerals and vitamins, the beet is an excellent addition to a healthy diet. The high natural sugar levels make it one of the sweeter-tasting root vegetables, and it has an earthy aftertaste.

You can eat the leaves and the roots of beets. Cook tender young beet greens as you would spinach.

RECIPE IDEAS

Beets are traditionally pickled for winter storage, but there are so many other ways to enjoy this superfood in both sweet and savory dishes. Raw beets are also delicious when grated and served in salads.

BEETS IN CHOCOLATE CAKE

Using beets in a cake may not seem like a good idea, but beets are actually very sweet and will make this delicious cake incredibly moist.

• Remove the leaves from 1 large beet and scrub clean. Put the beet in a pot of water, bring to a boil, and then cover and simmer for 1 to 2 hours until tender. Don't let the pot boil dry—add more boiling water if necessary. Let cool and then chop roughly.

• Preheat the oven to 375 degrees Fahrenheit (190 degrees Celsius).

• Put the chopped beet into a food processor and mix on high speed. Add a pinch of salt, 8 ounces (200 g) plain flour, 4 ounces (100 g) cocoa powder, 1 tablespoon baking powder, 9 ounces (250 g) superfine sugar, 3 eggs, and 2 teaspoons vanilla extract and then mix again on high speed until thoroughly combined.

• Gradually add 8 fluid ounces (200 ml) sunflower oil in a steady stream, as if you were making mayonnaise. Once it has all been added, stir in 4 ounces (100 g) dark chocolate pieces and pour into a loaf pan. Bake for an hour or until a skewer comes out clean.

• Serve as an afternoon treat with crème fraîche or whipped cream.

TASTY LAMB BURGERS WITH BEET RELISH

Burgers are all-time favorites that the whole family will enjoy. These lamb burgers are succulent when topped with a quick and nutritious relish.

• To make 4 burgers, finely dice 2 shallots and add half to a pan with a little oil and soften. Remove from the heat and let cool.

• In a separate bowl, place the rest of the diced shallots and grate in 2 large raw beets, a splash of olive oil, and a handful of chopped dill. Season, stir, and set aside.

• Place 18 ounces (500 g) ground lamb in a bowl and add the cooked shallots and a pinch of salt. Using your hands, knead the mixture well to combine and then shape into burgers. Put the burgers on a plate and chill for 30 minutes.

• Preheat your grill and cook the burgers for about 5 minutes on each side. Top with the beet relish and serve on toasted buns with a green salad on the side.

BEET AND POTATO SALAD

• Twist the leaves off 4 medium beets, scrub, and cook the beets whole in boiling water until tender but not too soft. Do the same with 4 medium potatoes.

• Let cool, cut the beets and potatoes into cubes of the same size, and gently stir them together in a bowl. Add 1 small red onion, finely diced.

• In a separate bowl, mix together 5 fluid ounces (150 ml) plain yogurt and a finely chopped green pepper and season. Stir most of the yogurt mixture into the potatoes and beets, keeping a little aside to spoon over the top.

• Garnish with a few parsley leaves and chill for 10 minutes before serving.

BEET, HORSERADISH, AND CRÈME FRAÎCHE DIP

Puree 2 medium cooked beets, 2 tablespoons horseradish sauce, and 4 fluid ounces (100 ml) crème fraîche in a food processor. Season well, spoon into a bowl, and serve with breadsticks, pita triangles, or vegetable sticks. Mixed up in minutes, and with a spicy horseradish kick, this dip is perfect for entertaining.

OTHER IDEAS

• Grate raw beets and add to a green salad, or make a simple beet salad with a squeeze of lemon juice, a pinch of ground cumin, and a handful of golden raisins.

• To pickle beets in jars, first prepare the jars by washing and rinsing them well and then sterilizing them by drying them at low heat in an oven. Use small beets for pickling, but don't cut them before cooking because the roots will "bleed" and lose color and nutrients. Boil until tender and then let cool slightly. Put as many beets as you can in each jar without damaging them and pour pickling vinegar over them. Seal the jars. Label the jars and store out of direct light until you want to use them.

3 LEGUMES

In a moderate climate, the most widely grown plants of the legume family are beans and peas, and one of the more practical aspects of growing these two vegetables is that they will grow happily together. Legumes gather nitrogen from the air and take it down through the plants into the roots, which means that your bean and pea plants are nourishing the soil for next year's crops.

Other everyday legumes include peanuts and lentils, but these tend to grow successfully only in hotter climates. If you have a greenhouse, however, or live in a hotter part of the world, there's no reason why you shouldn't try producing your own peanuts.

Beans

Beans make up a large group in the legume family, and they come in all shapes and sizes, so choose a variety that is suitable for your growing space and that you like to eat. Green beans and similar varieties are easy to grow in moderate to cooler climates, but they tend to be less hardy than peas. Although they need a warmer growing period, most like a lot of rain. The beans we shell, such as butter beans and lima beans, need a longer growing season, but if you start them early and protect them against cold nights, it's possible to grow these varieties in a cooler climate. Read the growing recommendations on the packets before you buy your seeds.

Green Beans
Seed

As mentioned, always check the variety's growing recommendations before you plant. With so many different bean varieties, it really is not worth trying to grow types that will struggle to thrive in your area. Because of their delicate root systems, most beans are generally sown in situ because they don't transplant well. However, if you are trying one of the longer-growing varieties, you could start the plants early, using biodegradable pots. Keep them in a warm place before planting them in their pots outside later in the year.

Different varieties of green beans also have different needs. Climbing varieties need a support system, which you should construct before you sow your seed.

Choose a sunny spot in the garden. Peas and beans grow well together in the same plot, so even though beans are usually planted later in the year, they can still occupy the same space.

Sowing

Dig over the ground and remove any perennial weeds, large stones, and nonorganic debris before raking to a fine consistency. Set up any support structures you need at this point. Low-growing bush varieties won't need any support at all, but taller-growing or climbing plants will need something. For example, a popular way to grow climbing beans is in a tepee design.

To create a tepee, push long canes firmly into the ground, in a circle. They should be about 6 feet (1.8 m) high when in place. Bring the tops of the canes together and tie firmly with garden twine to form a tepee shape. Attach strings around the canes every 8 to 10 inches (20 to 25 cm), starting from about 6 inches (15 cm) above the ground, and keep going until you get to the top of the tepee. Make sure the strings are taut and securely attached because your plants will use these for support. If you decide this design isn't suitable for your garden, you can train beans up a sunny fence or any other structure.

Beans come in many different varieties and types that thrive in many climates. Both green beans and fava beans (inset) do very well under moderate to cool conditions.

Keep your young bean plants warm with old window frames (shown here) or whatever you have handy.

a constant supply rather than having too many beans at once.

Care and Maintenance

Keep the young plants watered and weed free. Beans have a fairly delicate root structure, so be very gentle when hoeing. It's preferable to hand-weed where possible. Spray runner bean flowers with water twice a day, if the weather is dry, to "set" the flowers. If the plants are lacking water, the flowers will drop before the pods have formed. Other varieties will need watering regularly but not necessarily as often as runner beans. Check your seed packets for watering recommendations.

Harvesting

Pick your beans as soon as they are ready. Try not to leave them on the plant too long because they can become stringy (unless you are growing butter beans or another variety that you shell before eating). Generally, you leave shelled varieties on the plants until fully developed, but make sure that the autumn rains or cold nights aren't too harsh, or the plants will start to rot. Harvest early if you think the weather is about to turn nasty.

Eat green beans when they are small and tender for the best flavor and nutritional value. Green beans are particularly heavy-cropping plants, and you shouldn't deny

Sow beans in shallow trenches about 1 to 2 inches (2.5 to 5 cm) deep and about 2 to 3 inches (5 to 7.5 cm) apart. If you are planting bush varieties in lines, allow 6 to 12 inches (15 to 30 cm) between the seeds and 18 to 24 inches (45 to 60 cm) between the rows. Bush varieties are often very heavy-cropping plants, and it's a good idea to sow a line every two or three weeks through the spring and early summer for

yourself the treat of young beans. French beans (a longer, thinner type of green bean) should be at least an inch (2.5 centimeters) long before you eat them.

As with peas, leave the plants in the ground to completely die back. Then, you can either dig them into the soil to nourish the ground for next year's crops or cut the plants down to ground level and leave them to decompose into the soil.

Fava Beans
Seed

Fava beans are a good cool-climate crop and can be sown in either autumn or early spring. Many growers prefer to start fava beans in autumn because they are less likely to be attacked by black fly. Fava beans are especially vulnerable to this pest, although nipping off the tops of the plants sometimes acts as a deterrent.

If you are sowing your seed in the autumn, choose as sheltered and bright a spot as possible. Using a support structure is recommended because you can protect your plants from winter winds by tying them firmly to the support. Push two fairly long, strong canes or sticks into the ground at either end of the row and join them with a couple of crossbeams or some string. Use wooden slats or more canes or sticks and tie them firmly together. If the structure is weak, it can be counterproductive because

the wind may blow it down, taking your plants with it.

Sowing

Once you have your structure in place, sow seed about 2 inches (5 cm) apart. Check the manufacturer's sowing recommendations before you start. Sow about 1 to 2 inches (2.5 to 5 cm) deep and water well. Make sure that your plants get enough water, especially in the first few weeks. Fava beans grown through the winter months tend to get enough rainfall to thrive.

Care and Maintenance

Keep an eye on your plants until they are established and watch out for slug and snail attacks. If you are sowing in spring, your fava beans may not need support, but check your seed packets for any instructions or recommendations.

Harvesting

Pick your fava beans as soon as the beans inside the pods start to swell. The pods do grow fairly large and can sometimes be deceptive, so give the beans enough time to develop fully without leaving them on the plants to get too old. Young fava beans are a particularly delicious treat.

Fava beans are shelled before eating.

Storing Your Beans

Because there are many different types of beans, storage methods vary. You can freeze green beans, runner beans, and any complete pod beans. First, blanch the beans in boiling water for five minutes, drain, and cool completely. Freeze quickly to limit the loss of color, nutritional value, and taste. You can also can beans at home successfully if you have the proper equipment.

The beans that we shell before eating, such as butter beans, fava beans, and kidney beans, can be dried for storing, and they will store almost indefinitely in the right conditions. (**Note:** fava beans can also be frozen after blanching, and they will keep for several months in the freezer.) Shell the beans, remove any damaged ones, and blanch for five to ten minutes in boiling water. Remove from heat, drain, and cool. Spread the beans out in single layers on baking trays and heat in the oven on low heat for a few hours, turning them every half hour or so. When they are completely dry, allow them to cool and then store them in sealable jars out of direct light. Peas and beans tend to keep for a long time, so it is best to label them with dates so that you always use the oldest ones first.

Nutrients in Beans

Beans are a good source of protein, dietary fiber, calcium, and vitamins A and C. Always cook beans before eating because they contain a toxin that is harmful in their raw state. Once cooked, beans are perfectly safe to eat, of course. Cooked green beans can actually aid digestion and are safe to eat every day. After cooking beans, retain the water; it will contain iron from the beans, and you can use it to make gravy or sauces.

RECIPE IDEAS

Beans are very versatile in the kitchen and can be used in many different ways. Remember to store some for the winter months.

GREEN BEAN AND FETA SALAD

Remember to always cook beans before eating them; don't be tempted to throw them in the salad bowl raw, however harmless they look. They contain a toxin that needs to be cooked out first so that they are safe to eat.

• Place your beans in a saucepan and cover them with water. Bring to a boil, reduce the heat, and simmer until just cooked. Drain and cool completely.

• Shred a few lettuce leaves into a salad bowl and then stir in the beans. Crumble a little feta cheese and stir into the salad together with a handful of pitted black olives. Dress with vinaigrette or serve dressing separately.

• To make a simple dressing, put a tablespoon or two of white wine vinegar in a small bowl and add salt, pepper, and a spoonful of Dijon or other prepared mustard. Mix well. Just before serving, stir in a tablespoon or two of olive oil and whisk quickly with a fork or hand whisk.

• Serve immediately.

CREAMY GREEN BEANS

• Cook your beans, drain well, and cool. In a large wok or frying pan, heat a little oil or butter and gently cook about 3 ounces (85 g) flaked or sliced almonds (or substitute sliced mushrooms).

• Add the beans to the pan with a teaspoon or two of fresh chopped oregano. Stir well and cook gently for a couple of minutes until thoroughly heated.

• Stir in a few tablespoons of sour cream or crème fraîche, if desired. Remove from the heat and serve hot.

GREEN BEAN AND STILTON BAKE

• Preheat the oven to 400 degrees Fahrenheit (200 degrees Celsius). Blanch 9 ounces (250 g) green beans in boiling water for about 3 minutes. Drain well and transfer to a lightly buttered casserole dish.

• Pour 10 fluid ounces (284 ml) light cream over the green beans and then crumble 2 ounces (50 g) Stilton (or any other blue cheese).

• In a food processor, puree 2 slices whole wheat bread (crusts removed) to make bread crumbs. Mix the bread crumbs with 2 tablespoons melted butter and sprinkle evenly over the top.

• Cook in the oven for around 15 minutes or until browned.

FAVA BEAN PESTO

This rustic pesto recipe uses fava beans in place of the more traditional pine nuts and is delicious when tossed with warm pasta, potatoes, or fish. You can store this pesto in the refrigerator for up to a week.

- In a food processor, put 8 ounces (200 g) cooked fava beans (boiled for 10 minutes and gray outer skins removed), 2 garlic cloves, 1 large bunch fresh basil, 4 tablespoons grated Parmesan, juice and zest of half a lemon, 2 tablespoons olive oil, and salt and pepper.

- Puree ingredients until blended into a paste.

FAVA BEAN AND SHRIMP PILAF

- Finely chop a large onion and 2 garlic cloves. Fry in a pan with a little oil along with 1 teaspoon turmeric and 1 teaspoon ground cumin until the onion is soft.

- Stir in 8 ounces (200 g) basmati rice and cook for another minute. Add 1 pint (600 ml) water, season, and bring to a boil.

- Add 9 ounces (250 g) fava beans and simmer for 15 minutes. After this, add 9 ounces (250 g) shelled, cooked shrimp and simmer for another 5 minutes or so until the beans are tender and the rice has absorbed all of the water.

- Stir in a handful of chopped fresh cilantro and let stand for a couple of minutes. Serve with crusty bread or a green salad.

FAVA BEAN DIP

This tasty dip is perfect for a quick, healthy lunch or snack. Spread on crusty bread or dip with carrot sticks.

- Shell desired amount of fava beans, cover with water in a pot, and bring to a boil. Reduce heat and simmer until beans are tender. Drain well and allow to cool completely.

- Pour into a food processor and blend until smooth, mixing in a little olive oil to make the mixture smoother.

- Add a clove of crushed garlic and some chopped parsley and then season and blend again until smooth.

OTHER IDEAS

- Boil or steam fava beans; toss with butter, black pepper, or any seasoning you like; and serve as a side dish.

- Use green beans in soups and stir-fried dishes.

- Any type of beans grown for the seed part rather than the pod can make an excellent substitute for meat in lasagna, bolognese sauce, and chili. Cook the beans and then mash them into a paste to use in place of minced beef. Add a few finely chopped onions, along with herbs and spices, and form into patties for homemade organic veggie burgers.

Peas

Peas have been grown commercially for many years, but there is still nothing like homegrown peas collected from your own vegetable garden. There are different varieties, so you should check carefully before buying seeds.

Snow peas and sugar snap peas are eaten before the peas inside the pod develop, and the whole pod can be eaten raw or cooked. Everyday garden peas take a little longer to develop but are worth growing, and kids will love shelling the peas. Try a few different varieties if you have the space. You generally plant peas early in the year, so they will be out of the way by late summer.

Seed

Peas, like beans, are usually sown directly outside. It may be possible to start some varieties in individual biodegradable pots, but most peas are fairly hardy and are better sown in situ whenever possible.

While some varieties may have different requirements (check your seed packets), you will typically sow peas fairly early in the growing season. There are bush and vine varieties available, and both will grow well in moderate climates. The vine varieties produce tendrils that will cling to a fence or other support system. Build your structure before you plant to avoid any damage to the seeds, root system, or young plants.

For support, a length of chicken wire stretched between two posts that are firmly pushed into the ground works well for peas, although there are different types of pea netting and trellis structures that you can buy from garden suppliers. Bush varieties may need canes or pea sticks to help support them, particularly in windy areas. Pea sticks can be made from tree prunings, traditionally hazel. A thin branch with a couple of smaller branches leading from it makes a good pea support.

Sowing

Choose a bright, sunny spot in the garden and make sure that the ground is well drained. Dig over and remove any perennial weeds, large stones, or nonorganic debris and rake over to a fine consistency. Don't be tempted to add nitrates to your soil. Legumes catch nitrogen from the air and nourish the soil without you doing a thing.

Sow your seeds 1 to 2 inches (2.5 to 5 cm) deep, leaving about an inch (a few centimeters) between each one and about 2 feet (60 cm) between rows. Immediately after sowing, water well. Water regularly if the weather is particularly dry (although because peas are typically sown in the early spring, they tend to get enough rainwater). Keep weeds away, and the seeds should germinate within a couple of weeks.

Care and Maintenance

When the plants are still small, vine varieties will start to produce tendrils. Watch out for these and gently curl them around the support system you have in place. After

Peas will thrive in cool to moderate climates as long as they have a sunny spot in the garden.

Most pea varieties require some form of support—twigs, stakes, and netting all work well.

helping the first one on each plant, the others will find their own way, but keep an eye on them in case one or two get lost.

Don't mulch around pea plants because the stems can easily rot. This is a cool-climate crop that doesn't need too much pampering. You also can grow peas in containers, but the containers must be large enough to accommodate the pea plants and pea sticks (or whatever support system you use).

Harvesting

Harvest snow peas and sugar snap peas as soon as the pods are about an inch (a few centimeters) long. They will keep on cropping, so don't wait for them to grow any bigger. If you do miss a few, pull the strings off both sides of the pods and either slice them to add to a stir fry or eat them raw.

Collect regular garden varieties when the pods are obviously swollen with peas inside but, again, don't leave them for too long. One of the nicest things about growing your own vegetables is that you can afford to eat them at their best, so don't miss this one.

If you are saving peas to dry, leave them on the plants until the pods start to split. Although it may be tempting to leave the last ones that grow to dry, it's better to save a few of the earlier pods so they have longer to develop and dry naturally on the plant.

Allow the plants to die back on their own. Try not to pull them up while tidying the space. If you leave them to die back naturally and then dig them into the soil, you are replenishing the soil, and the area will be more fertile for next year's crops. Or, alternatively, cut the plants down to ground level and leave the roots to rot away into the soil.

Storing

You can freeze peas for up to a year. Freeze quickly in single layers, bag them, and label.

Dried split peas are delicious and filling when made into hearty winter soups and stews.

You can also can them if you have the right equipment. Probably the easiest method, though, is to store them as dried peas. Allow the pods to dry as much as possible on the plant and then collect them when the pods begin to split naturally. Be sure to harvest all of the pods before the autumn rains and cold nights arrive.

Some peas may have holes in them where worms ate through the pods. An easy way to separate them is to put a handful of peas into a bowl of cold water. The ones that rise to the top have airholes in them, which means that worms have burrowed into them.

To dry peas, shell them and then blanch for a few minutes in boiling water. Drain and allow them to cool completely. Lay the peas on baking sheets and heat them in an oven on its lowest temperature until completely dry, turning them every half an hour or so. When dry, allow them to cool completely and then store them in airtight jars, out of direct light. Peas dried this way will keep almost indefinitely.

Nutrients

Peas are a good source of vitamins A and C, which are both antioxidants and will help fight free radicals in the body. Niacin is also present, which helps with the absorption of iron. Peas are easy to digest and are often served as a side dish to those recovering from illness. They are also believed to protect the body against many aging conditions, so they may, in fact, help you live longer.

RECIPE IDEAS

PEA AND MINT SOUP

Traditionally, pea soup would bubble away on the stove for days, with more ingredients added when available. This recipe is simple and quick, and it makes a tasty dish for early summer or for later in the year. You can use dried peas instead of fresh, but soak dried peas overnight in water before using. They will take longer to cook than fresh green peas.

• Melt a little butter and add a finely diced onion to a pan. Cook until soft and then add about 2 pints (1 liter) vegetable or chicken stock and 2¼ pounds (about 1 kg) peas.

• Stir in a handful of chopped fresh mint and season to taste. Bring to a boil and then reduce heat and simmer for 30 to 40 minutes or until peas are tender.

• Remove from heat and allow to cool slightly. Blend in a blender or food processor until smooth and then return to the pan and very gently heat through, stirring constantly.

• Serve hot in individual bowls with a swirl of light cream and a couple of mint leaves to garnish.

BASIL, PEA, AND PANCETTA TART

This is a great recipe for feeding the whole family or for when you're having guests. Using a ready-made pastry shell makes this dish quick and easy to whip up, and you can prepare it the day before you want to serve it.

• Bring 10 fluid ounces (284 ml) of heavy cream to a boil and then remove from heat and add a small bunch of fresh basil. Leave to infuse for an hour.

• Blanch 4 ounces (100 g) fresh or frozen peas in a pan of boiling water for 1 minute, drain, and cool quickly under cold running water.

• Heat the grill and cook 6 slices pancetta until crisp and drain on paper towels.

• Heat the oven to 350 degrees Fahrenheit (180 degrees Celsius). Strain the cream into a bowl through a sieve and beat in 3 eggs plus 1 yolk, a handful of grated Parmesan, and seasonings.

• Tear the pancetta into small pieces and sprinkle into the pastry shell along with the peas. Pour in the egg and cream mixture. (You may have a little of the mixture left over, depending on the size of your pastry shell.)

• Bake for around 50 minutes or until set in the middle. You can serve this tart hot or cold, and it is delicious with Parmesan shavings on top.

LEMONY PEA RISOTTO

This risotto couldn't be simpler, and it's a quick dinner that you can make after a day at the office or working in the yard and garden.

• In a large saucepan, heat a little olive oil and stir in 8 ounces (200 g) Arborio (risotto) rice for a minute, stirring constantly. Gradually add a ladleful at a time of 28 fluid ounces (850 ml) hot vegetable stock until the rice is almost cooked and all of the stock is absorbed. This will take about 20 minutes.

• Add 2 ounces (50 g) peas (fresh or frozen), cook for another 5 minutes, and then remove from heat. Stir in a handful of grated Parmesan and the juice and zest of half a lemon and season.

• Serve immediately with lemon zest and Parmesan scattered on top.

STIR-FRIED BEEF WITH SNOW PEAS

One of the tastiest ways to serve snow peas is in a stir fry. Be careful not to cook them for more than a couple of minutes, though, or they will lose their color and crispness.

• Marinate 8 ounces (200 g) finely sliced rump steak with 1 tablespoon rice wine (or dry sherry), 1 teaspoon cornstarch, and 2 tablespoons oyster sauce for 20 minutes.

• In a wok, heat a little peanut oil and add 2 coarsely sliced spring onions and 1 teaspoon finely chopped fresh ginger. Stir-fry for a minute and then add the beef, stirring for another 2 minutes until browned. Set aside in a warm bowl. Add a little more oil to the wok and stir-fry the snow peas for 2 minutes along with a pinch of salt and sugar.

• Return the beef to the wok, toss everything together, and serve with steamed rice.

OTHER IDEAS

When the peas have swollen in the pods, collect them from the plants and get everyone to help with the shelling—it's a lot of fun! As you shell them, remove any damaged or worm-infested peas and put them, along with the shells, on the compost heap.

• Gently rinse newly shelled peas and then put them in a pot, cover them with water, and bring to a boil. Reduce heat and simmer until the peas are tender, usually 15 to 20 minutes, depending on the size of the peas. Add a mint leaf or two to the pot if you like. Or, after the peas have cooked, drain well, return to the pot, and stir in a tablespoon or two of butter and a few freshly chopped mint leaves—perfect with new potatoes and a Sunday roast.

• Chop snow peas and toss into a green salad. The crisp, fresh sweetness of the pea pods brings alive all of the other flavors in your salad. You can also slice snow peas lengthwise and mix them into stir-fried dishes.

• Cooked peas are versatile and can add bulk, color, flavor, and nutrition to many dishes.

• Try peas in Spanish omelets by including cooked peas with the other vegetables that you add to your egg mixture before cooking.

• Use cooked, completely cooled peas to make a rice salad: stir them into a dish of cold, cooked rice with finely chopped onion and a red pepper. Season and add a squeeze of lemon.

4 GREEN VEGETABLES

You can grow many leafy green vegetables right through the winter months. It is important to include leafy vegetables in our diets because they are full of vitamins that help prevent colds and the flu during the colder months of the year. And when you pick and eat leafy greens fresh, the goodness is still locked in, allowing you to eat the most nutritious foods possible. You can even pick brussels sprouts with little caps of snow on them, and they are really delicious straight from the garden (no matter what anyone says about disliking sprouts!).

As well as the green vegetables listed in this section, winter cabbage and spring greens are others you might like to try growing. Plan to sow a few lines of late seed after the summer crops are done to keep your garden producing food for as many months of the year as possible.

Kale

Kale is a hardy green vegetable that will put up with almost anything the weather throws at it. It grows well in northern climates and is a popular vegetable in many places. Kale is a primitive cabbage and has been cultivated for thousands of years, proving itself to be a useful winter crop. It is a biennial plant, producing foliage in the first year of growth and flower and seed in the second year.

A nutritious leafy green crop, kale is a delight to pick fresh from the garden during the winter months and even nicer to eat during this time, when many green summer vegetables are unavailable. As well as being a hardy and easy plant to grow, kale has become very popular due to its superfood qualities.

Seed

There are a number of types of kale, including the curly variety that is often found in supermarkets. It's also possible to buy an ornamental variety, which you can use as a garnish but not eat, so make sure you know which variety you are buying.

Kale is a brassica, like cabbage, and can be sown directly outside or started indoors. Seed should be sown either in trays or pots in early spring for summer cropping or outside in the summer for winter crops. Check on the seed packet for early and late sowing recommendations.

Sowing and Transplanting

Sow in well-drained trays or pots of fresh soil; keep fairly warm and water regularly. Seeds should germinate in ten to fourteen days.

Thin seedlings when they are large enough to handle, leaving the strongest one in each pot to grow. Make sure that the soil is damp before you start the thinning-out process.

Try to keep your kale plants growing in the pots until you've dug the last of the pea plants back into the soil. You don't have to plant the kale where the peas were, but the kale would benefit from the nitrates in the soil that were fixed there by the peas.

To transplant, loosen the top of the soil with a fork or spade, but don't dig again. Plant the kale plants, allowing about 18 inches (45 cm) between each plant and 18 inches (45 cm) between each row to give them enough space to develop. Water and keep weeds at bay, especially in the early stages of growing.

Kale seed can also be sown in situ. Because kale is a winter crop, you don't really need to sow it until summer. Sow it thinly in shallow trenches in a sunny, well-drained spot in the garden. Only dig over the soil if you didn't already work it earlier in the year. Keep the area weed free and water regularly. Thin out seedlings to maintain 18 inches (45 cm) between each plant.

Care and Maintenance

Kale is one of the more accommodating brassicas and will thrive in most soils, although it's always better if the soil is not too acidic. If you suspect that your soil may be on the acidic side, neutralize

Kale is a versatile and nutrient-packed leafy green vegetable.

All brassicas, including kale, are prone to caterpillar infestations. Check the undersides of leaves every day.

it with an organic product a month or so before planting.

Kale is a hardy plant that is fairly resistant to bugs and viruses. However, if you are growing early crops, watch out for cabbage white butterflies; as a brassica, kale is a prime target. Check the undersides of the leaves every day if you've spotted the cabbage white butterfly at all in your garden or yard. Scrape off any eggs or remove the whole leaf and compost it.

Make sure that your plants don't dry out in the summer months. As autumn

approaches, it's advisable in very windy areas to firm down the plants and earth up the stems to protect the roots from too much ground frost. Be sure to remove any yellowing leaves and keep the area free from weeds. Check your seed packet for other growing recommendations specific to your variety of kale.

Harvesting

You can use early-grown kale as soon as the leaves become available. If you cut the plants down in late summer, they may grow new shoots that you can collect in the winter months. There may be information on your seed packet about harvesting your specific variety of kale.

From around late autumn onward, start cutting kale leaves for use in the kitchen. Never use yellowing or older leaves because they will taste bitter. Take the first leaves from the crown of the plant, and only take a few leaves from each plant at a time because this will encourage the plant to produce new shoots from the main stem. The majority of your crop will come from these new shoots and will be ready to use from midwinter into early spring. Cut the leaves with a sharp knife or pull each leaf downward, steadying the rest of the plant as you pull.

Storing

Because kale crops right through the winter months, there is very little need to store it. It will keep for a couple of days in the produce compartment of the refrigerator, but it does tend to become more bitter in flavor the longer you keep it, so eat it fresh wherever possible. Kale is a cold-weather vegetable and will go limp very quickly in a warm kitchen. Also, don't wash the leaves before you put them in the fridge.

Nutrients

Kale is absolutely packed with nutrition. It is an excellent source of vitamin C and many of the B vitamins. Fresh kale also contains plenty of calcium and folate as well as dietary fiber. It is particularly high in potassium, with 228 mg per 4 ounces (100 g); this is a vital mineral for many important functions in the body.

The edible plants in the *Brassica* genus are known as cruciferous vegetables, which have been shown to reduce the risk of certain cancers. Research indicates that they may also help reduce the risk of heart disease—good reasons to make kale one of your recommended daily servings of vegetables.

With so many ways to incorporate this superfood into a healthy diet, more and more people are trying kale.

RECIPE IDEAS

Green vegetables often take a back seat when it comes to imaginative flair in the kitchen. It's all too easy to serve them on the side and then go through the routine of convincing everyone to "eat their greens." With a little creativity, though, you can jazz up kale so that everyone will want seconds. Not only do you get to feed the family tasty and nutritious food, you can also enjoy your own dinner without having to argue with the kids about finishing their vegetables! Try these ideas and invent a few of your own.

MEDITERRANEAN KALE

Lemon juice gives kale a distinct Mediterranean taste.

• Make a dressing with olive oil, a little salt and black pepper to taste, and freshly squeezed lemon juice.

• Rinse kale, cut into strips, and put into a large pan.

• Cover with water and bring to a boil. Reduce heat and simmer for 10 minutes or so until just cooked. Drain and then toss with the dressing.

• Add other ingredients that you have available—a clove or two of crushed garlic, some crumbled feta cheese, and halved olives will all add to the Mediterranean taste.

KALE PASTA WITH CHILI PEPPER AND ANCHOVY

• In a large pan, boil some penne or other tubular pasta. While it's cooking, heat 2 tablespoons oil in a saucepan and sauté a finely sliced red chili pepper, 2 chopped garlic cloves, and 4 finely chopped anchovies. Add shredded kale and then gently sauté until tender, adding a drop of water if needed.

• Drain the pasta, reserving a few tablespoons of the cooking water, and then toss the pasta and water with the kale, adding 2 tablespoons olive oil, the juice of half a lemon, and grated Parmesan.

• Sprinkle Parmesan shavings on top before serving.

KALE, PUMPKIN, AND BACON STEW

Make a tasty stew that will warm the family on a cold winter's day. There are endless recipes for stews, but this one is simple to make and has broad appeal; plus, you can use whatever ingredients you have available—potatoes, leeks, carrots, parsnips, and turnips, along with a chopped onion or two, are all tasty additions to this stew.

• Finely chop a package of bacon and fry it in a large pan until crisp. You shouldn't need any oil because the bacon fat will melt quickly.

• Peel and chop one medium pumpkin into small cubes and add to the pan with 3 halved shallots.

• Fry until the edges start to brown and then add enough chicken or vegetable stock to cover the ingredients.

• Bring to a simmer and then cover. Cook for 15 minutes or until the pumpkin is tender.

• Stir in 8 ounces (200 g) chopped kale, cover again, and cook for about 5 minutes or until tender.

• Stir in a handful of freshly chopped parsley and season well. Serve with toasted crusty bread.

OTHER IDEAS

• Serve boiled, steamed, stir-fried, or braised as an accompaniment to meat or fish dishes.

• Add to soups and casseroles to add bulk, flavor, and plenty of nutrition.

• Finely slice and use as filling for a quiche or calzone.

• Stir young leaves into winter salads.

• Add kale leaves to green juices and smoothies for a nutritious breakfast or boost of energy.

Brussels Sprouts

Sprouts are notoriously difficult to include in a family diet because there is always at least one person who declares passionately that he or she does not like sprouts. However, don't let that deter you. Sprouts grown in your own vegetable garden are very different from those you buy.

Brussels sprouts are a hardy winter vegetable and will grow happily in very cold temperatures. I once tried to grow them in France, but the winter wasn't cold enough for them to develop into hard sprouts. The plants grew well and produced tiny lettuce-like growths on the main stem, which were delicious. I managed to get the kids to eat them by telling them it was a variety of spinach!

Brussels sprouts seem to have originated in Europe and, although they are very defined miniature cabbages, they haven't been cultivated for much longer than a few hundred years in backyard gardens. Cabbages, on the other hand, have a much longer history and appear in records dating back thousands of years.

Seed

Sprouts are a long-growing, cool-climate vegetable, and you should sow seed around early to mid-spring. If the ground is prepared and not too cold, you can sow seed outside in situ. Plants tend to develop a healthier and stronger root system if they are planted in situ and do not suffer the trauma of being transplanted. However, if you sow the seed in biodegradable pots, it shouldn't make too much difference

overall. Plant the entire pot in the soil, and the pot will disintegrate into the ground. Whatever you choose, check the growing recommendations on your seed packet before you start to sow.

Sowing and Transplanting

Sow seed in well-drained trays, pots, or biodegradable pots, using fresh soil. Keep protected and fairly warm but not at hothouse temperatures. Water regularly. When seedlings are large enough to handle, thin out to leave one plant per pot in biodegradable pots; if you are transplanting later, you can leave two or three plants per pot.

If sowing seed directly outside, wait until all danger of frost has passed and plant in a well-prepared and well-drained bed in a sunny spot. As with all brassicas, sprouts don't thrive in an acidic soil, so make any adjustments to the soil in the month or two before sowing if necessary.

Sow seed thinly in shallow trenches. Water carefully and keep the area weed free. Thin the seedlings to allow about 4 inches (10 cm) of growing space per plant. After a few more weeks, thin again, removing weaker plants to allow about 18 inches (45 cm) of growing room.

Although brussels sprouts are a hardy crop, you shouldn't put young plants outside until all danger of frost has passed, usually around

Brussels sprouts may be one of the most underrated green vegetables, but there are many delicious ways to prepare and serve them.

Young, growing brussels sprouts.

Care and Maintenance

Make sure that your plants don't get bogged down with weeds. In late spring and early summer, weeds can take over in a few days, so keep an eye on your crops. Remove weeds by hand wherever possible because the sprout plants have shallow root systems and can be damaged by hoes or trowels.

Brussels sprouts are not immune to the cabbage white butterfly, so look out for this pest. Another problem from which sprout plants can suffer when they are very small and vulnerable is slug and snail attacks. These creatures will attack any young plants, and they can eat through the stems of a whole row of plants in one sitting, so watch out for them. However, sprouts overall tend to be pretty hardy plants and will resist most diseases.

Try not to let your plants dry out during a hot summer. If you find that the ground is too dry, shade the plants as much as you can during the hottest part of the day.

mid- to late spring in cooler climates. When your plants are 8 to 12 inches (20 to 30 cm) high, you can transplant them to the garden or to large containers outside. There's no need to overprotect sprout plants. Keeping them too warm is counterproductive because they prefer a cool growing climate.

It's a good idea to choose a fairly firm soil and put a stake in the ground just before planting. Brussels sprout plants are top-heavy and can get blown down in strong winds. Tie the plants to the stakes with garden twine as they get bigger, allowing room for the stems to grow. Allow 18 inches (45 cm) between plants and between rows and plant carefully. You can plant biodegradable pots without disturbing the plants, but water well to give the roots the chance to break through the pots.

Harvesting

A healthy plant in ideal conditions can produce around 2¼ pounds (1 kg) of sprouts. If you find that the sprouts don't form into complete, round, and hard little cabbages, don't worry. They are still edible, and you may be able to get away with convincing the "sprout haters" in the family that they are not even sprouts at all!

Pick sprouts as soon as they are ready. Pull each one off the stem or cut with a sharp knife. You can even cook and eat the top leaves of the plant as early spring greens. Sprouts will often crop from midwinter right through until early spring, depending on the variety and the weather conditions. In ideal conditions, the plant may develop more sprouts after you have picked some of them. Try not to strip a whole plant all at once, and you may be rewarded with a second crop.

Storing

Brussels sprouts are another cold-weather crop that can usually be left in the ground safely throughout the winter months, so you can pick them when you need them. They do, however, freeze quite successfully. Remove any damaged parts and then blanch in boiling water for five minutes. Drain well and cool completely. Put the sprouts into labeled containers or freezer bags and store in the freezer. Sprouts will keep in a cool place or in the produce drawer of the refrigerator for a few days.

Nutrients

Brussels sprouts contain the same nutrients as kale although in slightly different quantities. Sprouts are a good source of vitamins C and B as well as being high in potassium. They are also a good source of folate and calcium. Their vitamin content will help ward off winter ailments and boost the immune system. All cruciferous vegetables (edible brassicas) may help prevent certain cancers, and they generally improve your health and that of your family.

Heavily laden brussels sprouts plants ready for use; pick the individual sprouts as soon as they are ready.

RECIPE IDEAS

The secret to the best-tasting brussels sprouts is simple: do not overcook them. Sprouts actually release a sulfur-type chemical when overcooked, which makes them taste very bitter. Be imaginative in the way you prepare and serve your sprouts; for example, cook or dress them in olive oil or mix them with nuts or other green vegetables. You can chop or even mash cooked sprouts and mix them with other vegetables, including roots, such as potatoes or turnips. Add a few leftover sprouts to soups, stews, or casseroles. Use the liquid left over from cooking sprouts in soups and stews to give your dishes an extra boost of vitamins.

SPROUTS AND CHESTNUTS

Sprouts and chestnuts are a classic combination. Nuts complement green vegetables really well, and roasted or boiled chestnuts mixed with a bowl of steaming sprouts make something very special.

• Prepare your sprouts by removing any damaged or coarse outer leaves. Put in a pan and cover with water. Bring to a boil and then reduce the heat and simmer until the sprouts are just cooked.

• Drain well.

• At the same time or before, cook your chestnuts. Either peel and roast them or boil until tender.

• Stir the chestnuts and sprouts together in a serving dish and top with a tablespoon or two of butter. Serve hot.

BUBBLE AND SQUEAK

You can use any greens, including sprouts, to make this traditional British dish.

• Use leftover sprouts, if you have any; otherwise, cook the sprouts first and then chop finely and stir into a bowl of mashed potatoes. Add seasoning and fresh herbs to taste.

• Heat a little oil in a frying pan and then add the mixture to the pan. Cook gently so that a golden crust forms on both sides.

• If you prefer, you can form the mixture into individual-sized patties before cooking. Roll into balls and flatten slightly before putting them in the frying pan or bake them on a sheet in a preheated oven for 20 minutes or so until golden.

STIR FRY

Stir-fried sprouts taste delicious. Choose smaller ones if possible or cut larger ones in half.

• Boil the sprouts, drain well, and then add to the other stir-fry vegetables in your pan or wok with a splash of olive or peanut oil. (You can shred the sprouts instead before adding them to your stir fry, in which case you won't have to boil them beforehand.)

• Heat gently until completely cooked. Serve immediately.

OTHER IDEAS

• Steam brussels sprouts instead of boiling them.

• Fry cooked sprouts in a pan with pieces of salty pancetta or bacon.

• Add grated or chopped and cooked sprouts to mashed potatoes with a sprinkling of Parmesan for a twist.

• Brussels sprouts make a surprisingly good puree. Puree boiled or steamed sprouts in a food processor with a touch of light cream and then season well with pepper and salt and perhaps a little grated nutmeg.

Winter Lettuce

Winter lettuce is fun to grow and brings a touch of summer to cold-weather meals. Some of the Chinese leaf varieties, known as Chinese cabbage, Chinese leaf, or Chinese lettuce, can be planted late in the summer months for autumn picking. There are hundreds of different types of lettuce, so double-check that you are buying what you want. Check seed packets for variations in growing recommendations.

Winter lettuce is often very simple and straightforward to grow, but, sadly, it's often neglected as a winter crop. Although we don't tend to eat as many cold foods in the winter, there is nothing quite like the taste of a fresh green salad at any time of the year, and winter lettuce is often slightly sweeter tasting than the summer varieties. Unless you intend to eat a lot of winter salads, a few healthy plants will probably be all you need to see you through the cold months.

Seed

Choose your seed carefully. There are some lettuce varieties that are suitable for all-year growing, although they may need a little extra protection during the colder months of the year. It's a good idea to have a small cloche available to protect all young plants during cold weather, especially frosty nights.

Experiment with a couple of lettuce varieties if you have the space. The "cut-and-come-again" types are loose-leafed, and you can steal a few leaves from the plants every few days. 'Arctic King' and other winter lettuces can be grown successfully outside in cool climates, although they may need a little overnight protection during the coldest weather.

Sowing and Transplanting

Sow the seed from late summer to late autumn. If you are sowing directly outside, sow sooner rather than later to avoid frost damage to seedlings. Prepare well-drained pots or trays with fresh soil and sow seed thinly. Keep warm and watered.

Prepare a sunny, well-drained spot in the garden. Choose a sheltered area, if possible, to protect your plants from cold winds in the winter months. Dig over the ground lightly and incorporate some well-rotted compost to feed the soil if necessary. Transplant seedlings when they are large enough to handle; allow about 12 inches (30 cm) all around for the lettuce to fully develop. Double-check the recommendations on your seed packets because varieties differ in their spacing requirements.

Sow seed as thinly as possible in shallow trenches. Sow short lines of winter lettuce seed every couple of weeks or so from late summer to late autumn. Water well after sowing and then regularly until later in the year, when the rainfall will probably be enough to keep the plants watered. Thin out your plants as they start to crowd each other. Make sure the ground is wet before you start removing plants from the row, and remember to eat the ones you pull up.

Plant lettuce based on how much of it you plan to use during the winter.

Care and Maintenance

Although winter lettuce is bred for cold weather, a sharp drop in temperature may kill off seedlings and young plants. Keep an eye on the weather and protect them with a cloche or similar cover overnight. Remove the cover during the day, when the temperature has risen.

If you have to leave the cover over your lettuce plants for any length of time, remove it temporarily every day or so and tend to your plants. The soil under the plastic can become moldy and needs to be agitated from time to time to ensure that the lettuce roots don't start to rot. You may also need to water the plants and remove any weeds.

DID YOU KNOW?

In ancient times, lettuce was used in medicinal preparations. It can have mild sedative properties and was eaten in Roman times after meals to induce sleep.

If you are expecting a severe winter, try growing some winter lettuce in a cold frame or an unheated greenhouse. Don't forget to water plants if you use a greenhouse or cold frame.

You can even try to grow a head of lettuce indoors in the winter.

Harvesting

Winter lettuce crops can be used when the plants are big enough. Varieties that produce firm heads are best left to fully mature, although you can take a leaf from time to time. You can and should pick the cut-and-come-again varieties regularly. The plants will produce more leaves if you use them. Chinese leaf varieties should be left to grow until full maturity—if you can resist! Always check the growing recommendations for any extra tips on harvesting your chosen variety.

Storing

Winter lettuce is designed to grow throughout the winter months in relatively temperate climates, so it should be used fresh—straight from the garden. It will keep for a day or two in the produce drawer of the refrigerator if necessary. It will last a

Use lettuce straight from the garden whenever possible to get the best flavor and most nutrients.

little longer if dry, so don't wash the lettuce before storing. Close-leaf types of lettuce, like Chinese leaf varieties, will store for longer in the fridge, but you should still try to use them fresh from the garden whenever possible.

Nutrients

Fresh green vegetables from your garden have essential minerals and enzymes packed into their leaves, and daily portions will help boost the immune system and ward off illnesses. Lettuce, in particular, is a good source of vitamins A and C and folate. It is a very low-calorie food, making it easier to keep to a calorie-controlled diet, even in the winter months.

RECIPE IDEAS

We tend to think of lettuce as a summer vegetable and the main ingredient in tasty salads throughout the warm-weather months. But the more robust winter lettuces can also bring a touch of summertime to those dark winter evenings. You can serve virtually any dish with a side salad, but cooked lettuce is also a delicious option. You will be providing not only extra vitamins and minerals but also crisp textures, color, and, of course, extra taste, to your family meals.

LAMB AND LETTUCE PAN FRY

Because winter lettuces are slightly coarser than summer varieties, they are ideal to add to stir fries and pan fries.

• Heat a little butter and add 4 lamb neck fillets (or similar), cut into chunks. Cook for 6 to 7 minutes until browned on all sides and then add 2 handfuls of fresh or frozen peas.

• Add 5 fluid ounces (150 ml) chicken stock and simmer for a few minutes until the meat is cooked through.

• Add 3 good handfuls of chopped winter lettuce to the pan and simmer until it just starts to wilt but the color is still vibrant.

• Serve right away with boiled, buttered potatoes.

SALADS

Anything that you would normally put into a summer salad can be mixed into the winter salad bowl. Experiment with different combinations of vegetables according to what you have available. Shred lettuce leaves and put into a large bowl or on a serving dish with any or all of the following:

• Green peppers, cut into sticks. Cut a few thin circular slices across the whole pepper to use as a garnish.

• Finely sliced or chopped onion. Red onions can be slightly milder than white.

• Cherry tomatoes cut in halves or quarters or sliced larger tomatoes.

• A handful of mixed pumpkin or sunflower seeds.

• Halved walnuts, pine nuts, or other preferred nuts.

• Thinly sliced or grated raw carrots.

OTHER IDEAS

• Use in sandwiches and packed lunches—steal a couple of leaves or remember to use the cut-and-come-again varieties.

• Make a healthy and tasty lettuce soup.

• Use lettuce leaves to line a bowl and then spoon potato, beet, rice, or quinoa salad in the center.

5 HERBS

Herbs have been grown for medicinal, culinary, and cosmetic purposes for thousands of years. Many herbs are perennial, which means you can plant them and enjoy them for several years before they need replacing. Rosemary, for example, will live for up to fifteen years with a little care and attention along the way.

Using herbs in medicinal preparations should be done only with expert advice. However, there are many infusions you can make with herbs that will help prevent colds and flu and also gently ease minor digestive problems.

Herbs are a great addition to the vegetable patch because their strong scents help deter pests from your plants while at the same time attracting useful pollinators, such as bees, to your garden, especially when the herbs start flowering. Most will grow pretty much anywhere, such as in containers on a patio or in a sunny spot, and many will thrive on a bright windowsill. If you have the space, an herb garden is one of the easiest-to-maintain themed gardens; as well as being fragrant and beautiful to look at, it is also useful. Remember to use your herbs—they like to be picked regularly.

Sage

Sage, as its common name suggests, was considered the herb of wisdom in ancient times and was believed to bestow immortality. It has been used for centuries in medicinal preparations for various ailments and, although these days it tends to be used more frequently in the kitchen, recent studies have shown that a glass of sage tea every day can help reduce menopausal symptoms. Whether used as a medicinal or culinary herb, growing sage takes very little effort, and sage is a perfect addition to an herb garden. The variegated-leaf varieties will brighten up your garden even more.

Propagation
Seed

Growing sage from seed requires a little patience because the plants should be allowed to grow for a year before you use them. If you choose to try your hand at sowing seed, prepare well-drained trays or pots of fresh soil and sow seed in early spring. Keep pots inside or in a greenhouse and water regularly but don't overwater. Sage is native to the Mediterranean, so it isn't a thirsty plant, and it copes well in moderate heat. In the early stages of growth, it needs warmth. Plant outside later in the year after all danger of frost has passed.

Layering

Because sage is a woody plant, it's fairly simple to propagate new plants from layering the old one. Choose a healthy lower-growing branch from a well-established plant. Spread it on the ground and hold it down with a V-shaped peg (or a smooth stone or small rock) where it comfortably touches the soil. Cover with soil and then water.

If you do this in the autumn, and under the right conditions, your new plant may be

You can grow sage by sowing seed, layering, or taking cuttings.

ready for transplanting or replanting in the following spring. Carefully check that the layered branch has produced roots, remove the peg or stone, and then cut the stem from the mother plant. At this point, you can either leave your new plant to keep growing or transplant it.

Cuttings

You can also propagate sage from cuttings. Take 3- to 4-inch (7.5- to 10-cm) cuttings from a healthy plant and push, cut end down, into well-drained pots of fresh soil. Protect from the cold and water regularly, but, again, don't overwater. When the cuttings have produced roots, you can transplant them. If you don't want to wait for seeds, and you don't have access to a sage plant to take cuttings or layer from, buy a healthy plant from a garden center or other

Wait until the soil has warmed up and the danger of night frost has passed before transplanting young sage plants.

good garden supplier. If you grew the plants inside, don't put them outside too early— wait until the danger of frost has passed.

Transplanting

When the temperature of the soil has warmed up and the risk of frost has passed, put young sage plants outside, allowing enough space between them. Transplant into large containers (or pots, if you prefer) and keep in a fairly sunny spot. The most important element for sage to thrive is well-drained soil. Never let the ground become waterlogged. Sage does not cope well with very cold spells, so bring the plants indoors if you are facing a very severe winter.

However, I have had sage tolerate 6 inches (15 cm) of snow with no problems.

Care and Maintenance

You may find that your sage plants die back in the winter. Mulch the soil with grass clippings or similar organic material and remove early in the spring to allow the plant to start growing again. Sage plants will live happily for many years, although they can get straggly after four or five years. Cut back your plants to shape them, and take a few cuttings or layer in a few more plants to keep your supply going. Different varieties will have different needs. Some variegated sage grows in very convenient shapes and never seems to need pruning, but it doesn't always produce flowers. Flowering sage benefits from being cut back after the plant has finished flowering. Simply cut the dead wood back; this gives it a chance to rest and produce new growth.

Harvesting

All herbs like to be used. If you started your plants from seed, you should wait a year before picking the leaves. Otherwise, use sage as you need it, as soon as the stems are about 6 inches (15 cm) high. Pick leaves from different stems of the plants to allow them to regrow and produce more foliage. Stripping one branch puts pressure on a plant, and it may not thrive as it should.

Storing

You can often pick sage fresh all year round; however, you can also dry or freeze it if you expect a cold winter. To dry, hang sprigs of sage upside-down in paper bags so they don't get dusty. When the leaves are completely dry, crumble them into a jar and label the jar. To freeze, strip the leaves from the stems and freeze right away.

Nutrients

Sage contains significant amounts of vitamin C, calcium, and iron. It is an effective herb for reducing night sweats and other medical conditions. Years ago, sage was used to treat memory loss and digestive disorders. Its volatile oils are transformed into helpful medicinal aids, but the oils contain toxins so should not be taken in large doses.

When used properly, sage can have effective medicinal qualities.

RECIPE IDEAS

Sage is a strong-tasting herb and should be added to food only in moderation. Too much can be overpowering and can make the rest of the dish taste bitter. However, it complements chicken and other poultry beautifully, and sage stuffing is probably one of the most well-known chicken accompaniments.

SAGE TEA

An infusion of sage is not only a nutritious drink, but it also aids digestion and helps with various minor ailments. Simply pick a few healthy leaves, put them in a jug, fill with boiling water, and cover the jug. Let the leaves infuse for 5 to 10 minutes and then strain into a glass or small cup. Drink while still warm.

SAGE AND ONION STUFFING

Sage and onion stuffing has been available in boxed form for many years, but it's extra-tasty if homemade from scratch.

• Peel 4 medium-sized onions. Make sure they are crisp and have no damaged or soft parts. Put them in a pan and cover with water. Bring to a boil and then reduce the heat and simmer for about 10 minutes or until the onions are cooked.

• In the last couple of minutes of cooking time, add 10 sage leaves to the pan. Drain well and allow to cool for a few minutes.

• Chop onions and sage very finely in a large bowl. Stir in about 5 ounces (150 g) breadcrumbs and 2 ounces (50 g) softened butter. Mix together well and bind with an egg yolk. Add salt and pepper to taste.

• Shape into balls to make individual portions and bake on a baking sheet for 20 minutes or until slightly crisp.

BUTTERNUT SQUASH AND SAGE RISOTTO

Squash and sage together is a very tasty combination. This risotto has an autumnal feel, but it makes a great vegetable dinner at any time.

• Peel and cut a butternut squash into bite-sized chunks, toss in oil, add a handful of finely chopped sage, and roast in the oven for 30 minutes.

• While the squash is roasting, prepare the risotto. Bring 3 pints (1.5 liters) vegetable stock to a boil and keep it on a low simmer.

• In a separate pan, melt a tablespoon or two of butter and sweat a finely chopped onion for 8 to 10 minutes until soft and translucent.

• Add 12 ounces (300 g) Arborio rice and stir continuously for a couple of minutes until the edges of the grains become transparent.

• Pour in a small glass of white wine and simmer until evaporated. Add the stock, a ladleful at a time, and stir continuously until all the stock has been absorbed. After 25 to 30 minutes, the risotto will turn creamy, at which point you should remove it from the heat.

• Fry a small handful of whole sage leaves in a little oil and then drain on absorbent paper towels.

• When the squash is cooked, mash half of it into a puree and leave half as chunks. Stir the puree into the risotto along with a handful of grated Parmesan. Serve scattered with the chunks of squash and the crispy sage leaves.

OTHER IDEAS

Because sage complements roast chicken so well, consider adding it to other poultry dishes.

• Add leftover cooked chicken to stir-fry vegetables in a large pan or wok and then stir in a few very finely chopped sage leaves. Always make sure that reheated meat is piping hot all the way through before serving.

• Add a chopped sage leaf or two to chicken burgers or croquettes to give that Sunday-roast taste to midweek leftovers.

• Add 2 or 3 whole sage leaves to the roasting pan when cooking any poultry.

• Butter-fried sage leaves are a perfect companion to pumpkin ravioli.

Rosemary

All herbs have legends about magical qualities attached to them, and rosemary is no exception. Rosemary appears in records over thousands of years and was widely believed to improve the memory. Modern studies have shown that rosemary does, in fact, have an effect on the brain, so there may be reason to believe that the ancients knew what they were talking about!

Traditionally, rosemary is burned at weddings and funerals. It is associated with love and is often included in a bride's bouquet for luck. Mourners at funerals throw sprigs into the grave, bestowing loving memories of the deceased.

Propagation
Seed

Rosemary is not an easy herb to propagate from seed, but it can be done. Buy seed from a good supplier and check the seed packet for sowing recommendations before you start. Generally, seed should be sown in well-drained pots or trays of fresh soil early in the spring and kept indoors until later in the year.

Alternatively, you can sow seed directly outside after all danger of frost has passed. Choose a well-drained, sunny spot in the garden. Sow seed in shallow trenches and remove weaker plants when the seedlings are about an inch (a few centimeters) high. Allow them to keep growing and transplant if necessary the following year or when the plants are growing well and have developed a good root system.

Layering

As with sage and many other herbs, rosemary is a woody plant and propagates successfully by layering. Choose a well-established, healthy plant and layer the lowest branch or branches along the ground. Secure the branch with a V-shaped peg, a smooth stone, or a small rock; cover with a very light, sandy soil; and water. A new plant should grow from the stem and will, in the right conditions, develop roots. Cut from the "mother" plant when the new plant is growing well and has developed its own root system and then either allow it to grow where it is or transplant if necessary. Rosemary also does well in a pot, so consider moving your new plant to a container.

Cuttings

Take several 3- to 4-inch (7.5- to 10-cm) cuttings from a healthy, well-established plant after it has finished flowering for the year. Push the cuttings, cut end down, into a well-drained pot of fresh soil and then water it well. The soil should not be left to dry out completely but also should not be overwatered.

Keep tending to your cuttings until they have started to grow and develop roots, at which point you can transplant them into larger containers or directly into the garden.

Transplanting

Rosemary is originally a Mediterranean plant and thrives in well-drained soil; it should never be allowed to become

Rosemary's delicate flowers add a touch of beauty to the herb garden.

waterlogged. Shrubs tend to grow well near a wall or rock garden, and you should choose a sunny spot for them in your garden or grow them in large containers on a bright patio. A healthy plant can live for up to twenty years, so choose your spot well—it's far better for the plant if you don't have to move it later on.

When fully established, rosemary plants can grow up to about 4 feet (1.2 m) high, so position them near the back of an herb garden to avoid shading lower-growing plants. Rosemary is an evergreen plant and is a bright and colorful herb to include in the garden. If you have the space, plant a rosemary shrub in a corner of your vegetable patch.

Care and Maintenance

If you are expecting a very cold winter, keep new cuttings indoors in a greenhouse or in a sheltered spot until the weather warms up a little. Although rosemary is a hardy shrub once established outside, it will appreciate a little care and attention during the first year while it is developing.

Harvesting

Pick rosemary as you need it. Cut sprigs from healthy plants and try not to take too many stems from a plant at one time unless it is a well-established specimen. Cut 6- to 8-inch (15- to 20-cm) stems and then remove the leaves by running your fingers or a knife along the stems. Plants grown from seed should be left to develop for the first couple of years or so.

Increase rosemary by taking cuttings and placing them in a pot of well-drained soil, but don't overdo the watering.

Like many other evergreens, rosemary has needle-like leaves.

Storing

In most climates, rosemary will stay evergreen throughout the year and can be picked fresh when needed. However, if a very harsh winter is predicted, you can dry the leaves by hanging them in an airy place until completely dry. Use paper bags to prevent needles from dropping to the floor. Store the dried leaves in a jar out of direct light. It's easy to crumble dried leaves and add them to meals throughout the winter. You can also freeze sprigs, but because they are often evergreen and can be dried successfully, freezing seems unnecessary.

Nutrients

Rich in iron, calcium, and vitamin B6, rosemary is a useful herb, and although it isn't widely used in medicinal preparations these days, you can employ it in a number of ways. For example, add rosemary to bathwater to relieve tension: fill a small muslin bag with rosemary and hang it from the running faucet. You can also put a sprig under your pillow to encourage a good night's sleep. The essential oil extracted from rosemary is used widely in the cosmetic industry.

RECIPE IDEAS

Rosemary complements many dishes and has traditionally been used with pork, fish, and poultry; it has also become a popular ingredient in lamb recipes.

SALADS

The texture of the leaves doesn't particularly lend itself well to salads, although well-chopped or crumbled dried leaves can be stirred into a green salad. More often, rosemary is used to flavor oils and vinegars, which are in turn used to dress salads. Simply choose a healthy sprig or two from a well-established plant and put them into a bottle of white wine vinegar or olive oil. Seal the bottle and leave in a cool dark place for a couple of weeks to infuse.

MEAT DISHES

A sprig or two placed between the leg and breast of a chicken or turkey will help flavor the meat. Or you can strip the leaves from the stems and sprinkle the leaves into the cooking oil in the bottom of a roasting pan before adding the meat and/or vegetables. You can also add rosemary to stews, casseroles, and other one-pot meals.

Bouquet garni (a bunch of fresh herbs added to soups and stews during cooking and removed before serving) is a popular addition to stews and casseroles, and because there isn't any defined recipe, you can make your own from the herbs you prefer and have available. Rosemary, bay leaves, parsley, and thyme are all good herbs to add to the bunch. Simply tie some together with string and add them to the pot. Remember to remove before serving.

RED ONION AND ROSEMARY FOCACCIA

Focaccia is the perfect accompaniment to many Italian dishes and makes an excellent bread for dipping in sauces and olive oils.

• To make the dough, put 1 pound (500 g) bread flour, 1 packet (7 g) dried yeast, and a pinch of salt into a large bowl. Make a well in the center and pour in about 12 fluid ounces (350 ml) of lukewarm water. Use your fingers to combine the flour and water until you make a pillowy, workable dough, adding a splash more water as needed.

• Knead the dough on a lightly floured surface for at least 10 minutes until smooth and elastic.

• Place in a bowl, cover with plastic wrap, and allow to rise in a warm place until the dough has risen to double its size.

• When the dough has risen, punch it down and stretch it to fill a 9 x 13-inch (23 x 33-cm) cake pan. Let the dough proof for 20 minutes.

• Preheat the oven to 400 degrees Fahrenheit (200 degrees Celsius).

• Meanwhile, slice 2 large red onions and gather a handful of rosemary sprigs. Strip the leaves from the branches. Spread the onions over the dough and scatter the rosemary leaves. Press dimples into the dough with your fingers and

drizzle a generous amount of olive oil over the top along with a pinch of sea salt.

- Bake for 30 minutes until golden. When cool, cut into pieces; for a more rustic look, tear into pieces.

OTHER IDEAS

If you like the taste of rosemary, you can add it to almost any dish. It also has some functional uses around the kitchen and outdoors.

- Add a tablespoon or two of chopped leaves to apples when making apple jam or jelly.

- Use the stems as skewers. Strip the leaves and thread cubes of meat and vegetables onto the stems before grilling. The rosemary oil in the stems will add flavor to the meat and vegetables.

- Sprinkle a few leaves or a whole sprig on the barbecue coals to keep the mosquitoes and gnats away on summer nights.

- Put chopped rosemary leaves in the roasting pan along with oil when making roasted potatoes.

Thyme

Thyme has been used as a medicinal herb since ancient times. Roman soldiers were encouraged to bathe in thyme before going into battle because it was believed to bestow courage, and the ancient Greeks used thyme for embalming. Thyme leaves were also embroidered onto cherished linens and given as gifts to acknowledge bravery or adoration. As well as being used to flavor foods, thyme was often employed as a meat preservative. These days, thyme is grown for both culinary and medicinal purposes—it soothes sore throats and eases cold and flu symptoms. Native to the Mediterranean, thyme prefers a light, dry soil and will thrive on walls and in very well-drained soil.

Propagation
Seed

You can start thyme from seed quite easily, but you should leave it to grow and become established for a year or so before you begin taking its leaves. If you have the patience, it is certainly worth trying to propagate it from seed. You can sow seeds indoors in well-drained trays or pots of fresh soil and keep them warm and watered until later in the year. Transplant outside after all danger of frost has passed.

You can also sow thyme seed outside. To do so, place the seeds in shallow trenches in late spring—again, after all danger of frost has passed. Before you start, check your seed packet for growing advice to find out the specific requirements of your variety. Thyme grows successfully in pots and containers and won't mind a little drying out from time to time. It will need some water, though, so don't forget about your thyme completely.

Root Division

Thyme is a woody shrub and will benefit from being divided into smaller clumps every few years. The best times are probably

Unlike sage and rosemary, thyme is quite easy to start from seed.

either later in the year, after the plant has finished flowering, or very early in the spring, before it starts to grow. Try to avoid disturbing your plants when the ground is very cold, though. Dig carefully around your plant and divide it into two or more pieces. Gently but firmly pull the clump of roots apart and then replant immediately so the roots are not exposed for too long. Water well after transplanting and cover with a cloche or similar protection after planting if the winter is particularly harsh.

Cuttings

Start new plants by taking cuttings from a healthy and well-established plant. Cut 3- to 4-inch (7.5- to 10-cm) pieces of stem with a sharp knife or pruning shears. Push the cuttings, cut end down, into well-drained pots of fresh soil and water gently. Make sure that the soil stays damp but not too wet. Keep pots of cuttings indoors or in a greenhouse until they have developed roots and are starting to grow. They should be planted outside in early summer, when the ground has warmed up a little.

Transplanting

Thyme likes a light soil in a sunny spot and will thrive near walls or rock gardens or in raised beds. It is possible to grow in heavier soil, but it loses some intensity of taste and fragrance. Thyme is also a very suitable container herb, especially because it doesn't

You can plant thyme in open ground, but it is also equally happy in pots and containers.

mind if you forget to water it for a while. Plant it in a container on the patio or in any sunny spot in your yard or garden. It will also do well in pots or containers indoors as long as it has enough light.

Care and Maintenance

You must never allow your thyme to become waterlogged. In fact, it will survive longer without any water than with too much, so resist overwatering.

Generally, thyme will stay green throughout the winter months, making it an ideal winter herb to grow. If your plant

You can enjoy thyme's medicinal benefits by brewing the herb into a tea.

does die back, or the temperature gets very low, mulching around the roots will help protect it. Remember to remove the mulch before the following spring, if it hasn't decomposed, to allow the plant to grow.

Harvesting

Leave any plants you have propagated from seed to grow for the first year so that they develop strong root structures. After that, or with older plants, pick sprigs as you need them. As with most herbs, thyme likes to be picked regularly and will produce more foliage if you use it. After flowering, plant growth tends to slow down for the year. If you are expecting a very cold winter and have young thyme plants, collect some leaves to dry or freeze—just in case.

Storing

Thyme will often stay green throughout the year, and you should pick it fresh whenever possible for the full benefits of taste and nutrition. However, you can also successfully dry or freeze it. To dry, lay the stems or the leaves (separated) on a tray in a dry place until they are dry enough to crumble. With stems, you also have the option of hanging them in paper bags in a cool, airy place until dry. Crumble leaves into a jar, label, and store out of direct light. If you prefer to freeze them, lay the stems on a freezer tray and freeze quickly after picking. Put leaves in a suitable container and label before storing in the freezer.

Nutrients

A good source of iron, thyme is also a wonderful healing herb. It has natural antiseptic properties and is a quick remedy for cuts and grazes from the garden. Its natural oil, thymol, is considered to be a valuable ingredient in many commercial medications.

RECIPE IDEAS

Thyme is a useful herb to have in the kitchen, and it enhances the taste of many fish and meat dishes. It is popular in many countries, and a sprig of thyme is often included in *bouquet garni*. Thyme is also one of the ingredients in *herbes de Provence*, a mixture of herbs often available in dried form but much tastier when fresh. You should note that many recipes measure thyme by the tablespoon; in these cases, remove the leaves from the stems first. Also, thyme releases its flavors slowly and should be added, in most cases, early on in the cooking time.

THYME TEA

To soothe cold and flu symptoms, put a handful of thyme leaves in a jug and pour over boiling water. Cover and let stand for five minutes and then strain into a glass or cup with a dash of lemon juice and a teaspoon of clear honey. Sip while still warm.

EASY TOMATO AND THYME COD

When you would rather be on the couch than in the kitchen, this quick dinner couldn't be simpler.

• Heat a little oil in a pan and fry a finely chopped onion until soft.

• Stir in a 14-ounce (400-g) can of chopped tomatoes, 1 teaspoon brown sugar, a small handful of thyme leaves, and a good splash of soy sauce. Bring to a boil and then simmer for 5 minutes.

• Gently place the cod into the sauce, cover, and simmer gently until the fish is cooked, about 8 to 10 minutes. Serve with potatoes or crusty bread.

LEMON AND THYME TOGETHER

• Add a sprig of thyme to a pitcher of lemonade or another lemon-based drink.

• For a taste of summer, stir a few fresh thyme leaves into your salad bowl with a squeeze of lemon or make a dressing that includes finely chopped fresh thyme leaves.

• Add thyme leaves to a chicken stir fry early in the cooking time. Stir in a little freshly squeezed lemon juice just before serving.

OTHER IDEAS

• Add whole sprigs to the roasting pan when roasting meat or vegetables, or remove the leaves from the stems and sprinkle them into the bottom of the pan.

• Sprinkle a few leaves onto halved tomatoes before "sun-drying" them in the oven. Arrange tomatoes on a baking tray and sprinkle with thyme. Set the oven to the lowest temperature and remove the tomatoes once they are dry. Check the tomatoes during drying to make sure that they do not start to "cook." Cool completely and then store in containers until you use them.

• Try stirring some tiny thyme leaves (chop if necessary) into soft cheese before spreading on sandwiches or mixing with warm pasta.

Garlic

No good cook should be without onions or garlic, and garlic is, fortunately, fairly straightforward to grow. It stores well for most of the winter months and adds depth of flavor to hearty winter soups, stews, and casserole dishes.

Garlic is steeped in legend and myth, including the idea that it can ward off evil spirits. Romans believed it to be an herb of strength, and soldiers were encouraged to eat it daily. Its strong scent will deter most pests from your garden and will indeed protect your crops. Garlic's antiseptic qualities may also help ward off some viruses and diseases in humans.

Seed

Garlic seeds are simply the cloves from the garlic bulb. It is possible to grow further bulbs from cloves that you remove from supermarket-bought garlic; however, for the best crops, buy hybrid bulbs from a reputable garden supplier because they have been bred specifically for growing at home. If you do decide to try growing from a supermarket bulb rather than a hybrid, at least make sure that the garlic has been grown organically.

There are many different hybrid varieties of garlic available, from 'Red Sicilian' to elephant garlic, mostly originating from France. Some are suitable for autumn planting, and you should check for any recommendations before you buy bulbs. Autumn-planted bulbs may need a little protection during the colder months.

If you have the space available, try a couple of different varieties of garlic so you know what grows best in your garden for the following year.

Transplanting

Garlic is a hardy plant and will tolerate low temperatures. You can plant it early in the year, although some varieties can also be planted in autumn. For spring sowings, dig over the ground during the previous autumn if you can or as soon as the soil is workable in early spring. Clear out any perennial weeds and large stones.

It's important to remember not to overfertilize the soil when growing garlic. Overfeeding will encourage the plant to develop more leaf, which means that it won't put as much energy into producing the desirable swollen garlic bulb. Make sure the soil is well drained and rake the ground over to a fairly fine consistency. Plant the garlic cloves about 8 inches (20 cm) apart,

Garlic is a staple ingredient in cuisines around the world.

leaving 12 inches (30 cm) or more between rows so that you have enough room to maintain and weed your plants. Plant base end down and leave the tip of each clove just showing above the ground.

Garlic is also suitable for container growing because each plant takes up little space and doesn't need a great depth of soil. Don't forget to water them, though. The soil in containers and pots dries out quickly, and garlic won't thrive if it doesn't get enough water. Again, make sure that the pots are well drained and in no danger of becoming waterlogged.

Care and Maintenance

If you find that birds are a problem, you can cover the tips of the cloves lightly with soil or put a protective wildlife-friendly net over them until they put down roots. Once established, the cloves won't pull out of the ground easily. Also remember to water gently after planting. Garlic is generally sown in early spring, so it should get enough rainwater, but keep an eye on the soil to make sure that it does not dry out. Pull out any stray weeds to give your plants plenty of space to develop.

Garlic is a hardy plant, and, because of its strong scent, it does not tend to attract predators. The only thing that can be a problem with garlic is "bolting," or running to seed too quickly. Many plants bolt in hot, dry weather, but you can discourage this by shading them a little when the days are long and hot. Water regularly if there isn't sufficient rainfall to keep the soil damp. If

Garlic braids are both practical and attractive.

a garlic plant does bolt—it will send out a hard stem from the middle of the bulb— fold over the leaves or snap the hard, woody stem off. Such a bulb won't store as well as the others, so you should use it first.

Harvesting

Garlic behaves the same way as onions and shallots. It will grow until the bulb has developed to its full size, and then the leaves will start to droop and dry out. Some growers fold the leaves down when they start to dry, while others just leave them. If you feel like tidying up the vegetable patch, simply fold the leaves over and then leave them to die back completely.

Before the weather changes in the autumn, lift all garlic. Gently loosen the soil around each bulb, if necessary, and pull them out of the ground. Lay on dry earth or trays for an hour or two to dry out before brushing off any excess soil and storing.

Storing

The best way to store garlic is in braids if you can. It's not only practical, but it also decorates the kitchen nicely. Simply braid the leaves together when they are dry but not crumbly. This takes a little practice but is well worth the effort. If braiding isn't the best option for you, remove the leaves by cutting them about an inch (2.5 cm) or so above each bulb and store the bulbs in single layers in cardboard boxes or wooden trays. Keep them out of direct light in a cool, dry place.

Nutrients

Rich in vitamins B and C, calcium, and iron, garlic also has antibacterial, antiviral, and antifungal properties. It is an excellent addition to your diet in the winter months to boost immunity against colds and viruses. It is also said that garlic cleanses the blood, which could, of course, be the basis of the myths about warding off evil spirits. Those affected by eczema or other skin problems should consume garlic in smaller doses because garlic can irritate certain skin conditions.

Along with adding delicious flavor to your meals, garlic can help you ward off cold and flu symptoms.

RECIPE IDEAS

No good chef would be without a clove or two of garlic (and a few onions) in the kitchen. And a less-than-perfect cook will be able to hide all sorts of culinary hiccups with a crushed clove in the recipe! Chefs around the world use garlic in their cuisines, from Chinese to French to Mediterranean to Russian.

GARLIC BREAD

Garlic bread goes with many family dishes, such as spaghetti with meat sauce, chili, soups, and salads. Making garlic bread is an excellent way to use up day-old French bread.

• Using a sharp knife or a bread knife, cut slices along the loaf about an inch (a few centimeters) apart but do not cut all the way through (so that the loaf stays together).

• Blend a couple of cloves of garlic with slightly softened butter or butter substitute and stir in some finely chopped fresh parsley. Using a knife, divide the garlic-herb butter between the slices. Spread any leftover butter on top of the loaf.

• Wrap the loaf in foil and place on a baking sheet. Put in a preheated oven (medium heat) for about 20 minutes or until hot all the way through. Unwrap and serve hot.

STUFFED GARLIC MUSHROOMS

These mushrooms make a great appetizer or a tasty light lunch.

• In a pan, sauté 2 finely chopped shallots and 2 garlic cloves until soft.

• Mix with 9 ounces (250 g) bread crumbs, a little chopped parsley if available, and a pinch of salt and pepper.

• Wash and stem 12 large mushrooms and place in a greased baking pan.

• Pack the bread crumb mixture into each mushroom cap and top each one with grated mozzarella cheese.

• Bake in an oven on medium heat for 10 to 15 minutes or until the cheese is melted and golden and the mushrooms are fully cooked.

OTHER IDEAS

• Toss cooked, drained carrots with finely chopped garlic and a tablespoon of butter before serving.

• To give just a hint of garlic to a salad, rub a halved clove of garlic around the inside of the bowl before filling with salad ingredients. For more of a garlic kick, finely chop a clove or two and add it to your salad dressing.

• Rub a halved clove of garlic on slices of toasted French bread and then drizzle the bread with olive oil to make delicious bruschetta. Top with tomatoes for a real treat.

Chives

Chives are part of the same family of plants as onions, shallots, and garlic. They have similar properties to onions but are milder in taste and are a useful herb to grow in your backyard garden, in containers, or even on your kitchen windowsill. They will stay fresh and green all year round on a sunny windowsill as long as you remember to water them occasionally. If planted near vegetables, chives will help deter pests. Chives are fairly hardy perennials, which means that one plant should last for many years in the right conditions.

Propagation
Seed

There are different chive seeds available, including a "garlic" variety. Some varieties produce white flowers and others purple, so grow a few of each to enhance your herb garden with different-colored blooms. Chive flowers, as well as the leaves, are edible.

Chives are among the few plants in the onion family that will grow easily from seed. The germination rate is usually fairly quick, so chives are great plants for beginners—especially children, who like to see rapid results. In fact, there are special pots that are designed to grow chives as "hair" over smiley-faced pots, which children will enjoy. Alternatively, start seed off in early spring by sowing thinly in well-drained trays or pots of fresh soil. Check the sowing recommendations on your seed packet for regional variations before you start.

Keep your chive seeds warm and watered—but don't overwater—until the seeds have germinated. Don't let the seedlings dry out.

Root Division

Dig up a healthy chive plant that has been established for a year or two. To avoid damaging the roots, dig carefully around the plant when the ground is wet, gently but firmly separate the roots into two or more clumps, and replant immediately. Water well immediately after planting.

Transplanting

After all danger of frost has passed, you can plant chives in the garden. As long as the ground is well drained and in a reasonably sunny spot, chive plants can be positioned around the garden—they make good border plants because they don't grow very tall and are attractive to look at. Plant one or two chives in the corners of your vegetable garden to help deter bugs and attract bees.

Handle the seedlings carefully and allow space for them to grow. Although they like a sunny spot, chives may suffer if exposed to long, hot, dry periods. Make sure that they get enough water and rig up a little shade for them if the weather is scorching.

Care and Maintenance

Chives are rather hardy plants and will thrive in less-than-perfect soil. However, in poor

Among their many uses, chives are a natural pest deterrent in the vegetable garden.

soil, they will benefit from organic feedings from time to time. Chives are rarely affected by bugs and viruses, but to keep them healthy and producing fresh, green foliage every year, you should separate them by root division every few years. There are no hard-and-fast rules as to how often you should divide them, so it's a good idea to check the plants for any signs of rotting in the center or to see if they are simply getting too big for their allocated space. In a very cold winter, plants may die back but should revive in the spring. In a severe winter, mulch will protect the roots, but remember to remove any matter that has not decomposed before spring.

Harvesting

With a little luck and attention, a chive plant will produce fresh, green leaves practically all year round. Always use sharp scissors and cut leaves from the outside of the plant. Cutting from the middle restricts

the cut leaves from growing again because they have less light, but the outer leaves regrow easily. Cut to about 2 to 3 inches (5 to 7.5 cm) from the ground, depending on the size of your plant.

You can collect the flowers as soon as they appear; cut down the flower stalks when removing the flowers. After the plant has finished flowering, cut down the whole plant to 2 to 3 inches (5 to 7.5 cm) high, and it should start producing leaves again. Use plants grown from seed when they are growing well and have many leaves.

Storing

Because chives will stay evergreen for most of the year, you probably will not need to store them. However, if the winters are severe in your area, you can freeze fresh chives quickly after picking and store them for several months. Either place them on trays to freeze and then put them into suitable containers and label, or cut them into small pieces and put them in ice cube trays before filling with water and freezing. Simply defrost the ice cubes to retrieve the chives.

Nutrients

Chives contain similar nutrients as other members of the onion family, although their overall constituents tend to be weaker than those of onions or garlic. Chives are useful for culinary purposes, but they also help alleviate colds and flu symptoms and are considered to be an effective digestive aid. They have significant amounts of vitamin C, calcium, and iron with hardly any calories or fat.

Chives impart a delicate onion-like flavor to foods.

RECIPE IDEAS

Chives can brighten up any dish with a hint of onion flavor. They are perfect to add to dishes when you want a milder taste than that of stronger-tasting onions or shallots.

SALADS

Snip chives into small pieces with scissors or chop with a sharp knife and stir into salad greens. Chives will pep up any plain green salad, and they are particularly tasty when snipped over a ripe tomato salad dressed with olive oil and lemon.

POTATO SALAD

Chives bring out the taste of potatoes like no other herb.

• Peel and wash potatoes and then cut into cubes. Put in a pot and cover with water. Bring to a boil and then reduce heat and simmer until potatoes are just cooked. Don't let them get too soft.

• Drain well and let them cool completely.

• Make a dressing with equal parts mayonnaise and plain yogurt. You can use crème fraîche instead of mayonnaise or yogurt. Add a little cream if calories aren't an issue. Stir in chopped chives to flavor the dressing.

• When the potatoes are completely cold, transfer them to a serving dish and gently stir in the dressing.

• Chill in the refrigerator for at least 10 minutes before serving. Garnish with a few chopped chives.

OTHER IDEAS

• Sprinkle chopped fresh chives over cheese on toast and then heat under the broiler for a minute or two.

• Add finely chopped chives to scrambled egg, omelet, or quiche mixtures before cooking.

• Add small pieces of fresh chives to sandwiches; they are particularly good with tuna or egg-salad sandwiches.

• Sprinkle some chives over boiled new potatoes or mix some chives with sour cream and top piping-hot baked potatoes.

• Nibble on a few chives before dinner instead of diving into the bread basket. Chives will take the edge off your hunger and prepare your taste buds for savory foods.

Parsley

Parsley is probably one of the most popular and common herbs, although its typical role as a garnish means that it often stays on the plate rather than being eaten. Recent studies have shown that parsley is one of the most nutritious herbs we can grow—packed with vitamins, it is an excellent herb to add to many dishes. Parsley's breath-freshening qualities also make it an ideal herb to serve with garlic dishes.

Originating in the Mediterranean region, parsley is widely cultivated throughout the world. It is a biennial plant, so it produces leaves in the first year of growth and flowers and seed in its second year.

Seed

There are a number of parsley varieties that will grow readily in your herb garden; the most popular are the curly-leafed variety (often used as a garnish) and a flat-leafed Italian variety. Both are equally tasty and nutritious. Sow parsley every year for a constant supply but leave the plants in the ground if you can. Seed can take up to six weeks to germinate, and you will need to keep pots and trays warm and watered regularly. Some growers soak seed in water for a few hours before sowing to speed up the process.

In early spring, prepare well-drained pots of fresh soil and sow a few seeds in each. Check the seed packet for regional variations in growing recommendations. Keep pots in a warm, bright place. Make sure that your parsley never becomes waterlogged, but be aware that the soil should always be damp; otherwise, the seeds won't germinate. When the seedlings are about an inch (a few centimeters) high, carefully remove weaker plants from the pots and leave the strongest plant in each pot to keep growing.

In some areas, you can sow parsley directly outside, but unless you have a long growing season, you may find that the plants don't develop as well as those started earlier in a greenhouse. If you do decide to sow seed outside, wait until all danger of frost has passed and sow thinly in shallow trenches. Remove weaker plants when the seedlings are about an inch (a few centimeters) high to give the others space to grow.

Transplanting

Parsley is a very nutritious herb, but because it draws its minerals and vitamins from the soil, it is a heavy-feeding plant that will not grow well in poor soil. Choose a sunny, well-drained spot in the garden and dig in plenty of aged compost or well-rotted manure a month or two before growing, if necessary. Dig over the ground, removing any perennial weeds, nonorganic debris, and large stones, and rake over to a fine consistency. Before removing plants from their pots, soak them well in water so that you'll be able to pull the plants out without damaging the roots. You can use biodegradable pots if you prefer, but you should soak them in water

Parsley is a popular garnish, but you miss out on its many benefits if you leave it on the plate.

before transplanting. When transplanting parsley plants, allow plenty of space for them to grow; check your seed packet for any spacing recommendations. Water well after planting.

Care and Maintenance

Water your parsley regularly in dry weather and make sure that the plants don't dry out. Keep the soil around the plants free of weeds. In very long, hot, dry periods, the plants may appreciate a little shade from the midday sun. Parsley plants will benefit from organic feedings every week or two, especially if you've previously used the soil for other crops, because the soil's nutrients may be depleted.

Harvesting

You can start using parsley when the plants are about 8 inches (20 cm) high. This gives them adequate time to develop their roots. Pick stems as needed but don't strip a plant of all of its leaves; instead, take a few from each plant to allow the plants to grow more foliage. When the plant has obviously stopped producing leaves, cut a few sprigs for storing.

Parsley needs good soil to thrive, and it will benefit from regular feeding.

Leave the plants in the ground to overwinter. If conditions are good, they will produce flowers and then seeds in the following spring, which you can then collect for sowing in subsequent years or for use in the kitchen to flavor food.

Storing

Parsley will keep for a few days in your refrigerator's produce drawer but, like many herbs, is best when used fresh. You can also store parsley in the freezer: place whole stems on a suitable tray quickly after picking, freeze them, and then enclose them in labeled containers or freezer bags. Parsley is also easy to dry by hanging stems in a dark, dry, and airy place. Place cardboard underneath or place the stems in paper bags to collect any falling leaves. When completely dry, crumble the leaves into jars and label. If stored out of direct light, dried parsley will retain its flavor for many months.

Nutrients

Ounce for ounce, parsley has more vitamin C than citrus fruits. Logically, you would have to eat a huge amount of parsley to get the same amount of vitamin C as you would from an orange, but it is a worthy herb to add to any dish. Parsley is also packed with iron and other vitamins and minerals. It contains significant amounts of calcium and very few calories. As mentioned, parsley is a very effective breath freshener and should accompany strong-tasting dishes.

Parsley's small green leaves are big on vitamin C as well as other important nutrients.

RECIPE IDEAS

It's such a shame that parsley is more often used simply as a garnish rather than added to recipes. By ignoring the parsley, we could be leaving the most nutritious part of the meal uneaten. Parsley root is also edible and can be scrubbed, sliced, and added to casseroles, soups, and stews.

PARSLEY PIE

Parsley pie is an old Cornish recipe and is a great way to use leftovers.

• Half-fill an oven-safe dish with chopped fresh parsley and then stir in any leftover meat (traditionally lamb) and vegetables; you can also use vegetables only.

• Hard-boil a couple of eggs and add them to the dish.

• Pour vegetable or chicken stock almost up to the level of the parsley mixture.

• Roll out enough ready-made pie crust to fit the top of your dish and cover the parsley mixture with the pie crust. If desired, glaze the pie crust with a little milk or beaten egg before baking.

• Bake in a preheated 350-degree Fahrenheit (180-degree Celsius) oven for about 30 minutes. If you are using meat, make sure it is piping hot all the way through before serving.

• Serve hot with mashed potatoes or green vegetables.

PERFECT PARSLEY AND CHICKPEA SALAD

• In a large bowl, combine a 14-ounce (400-g) can of drained chickpeas, a finely sliced red onion, half a chopped cucumber, 4 ounces (100 g) crumbled feta cheese, and a good-sized bunch of roughly chopped parsley.

• Mix 2 tablespoons red wine vinegar with 3 tablespoons olive oil and season. Pour over the salad and toss to combine all ingredients.

OTHER IDEAS

• Use parsley in stews, one-pot meals, and tomato sauces by simply adding some chopped fresh leaves.

• Combine chopped parsley with bulgur wheat, spring onions, mint leaves, lemon juice, and olive oil to make tabbouleh, a traditional Middle Eastern salad.

• Make a rub for meat or fish by mixing chopped parsley with garlic and lemon zest.

• Add a combination of chopped parsley and mint to any green salad.

• Parsley sauce is a traditional accompaniment to fava beans and boiled ham. It is also delicious with any fresh garden vegetables, particularly beans and peas, and it provides a great savory base for a creamy chicken pie.

6 FRUITS

There are many fruits that we can grow in our yards and gardens, making it both delicious and easy to get the recommended daily servings of fruit that we know we should be eating. Harder fruits, such as apples and pears, thrive in cooler climates, and many of us have fond memories of the apple tree in a neighbor's yard. Nowadays, with varieties grafted onto solid rootstock, we can grow apples and pears in smaller spaces and still produce a lot of fresh fruit for the family.

We can also grow many soft fruits, such as strawberries, raspberries, and blackberries, in most climates. Traditionally a summer fruit, strawberries now come in varieties that produce fruit early and later in the year. They are also a perfect jam ingredient to bring a touch of summer to the winter months. The more exotic fruits, such as kiwis, bananas, and melons, need a long, hot growing season, so, unless you are set up with a heated greenhouse, these fruits aren't really suitable to grow in a cooler climate. However, every year seems to bring new hybrids onto the market that will cope with shorter, cooler growing seasons, so it's worth keeping an eye out for new fruits that you may be able to grow in your region.

Apples and Pears

If you have an average-sized yard, you may think that it's not practical to grow apple and pear trees, but you don't need a big space to grow these wonderful fruits. Grafted trees, available at good garden or tree suppliers, are basically varieties of smaller trees grafted onto hardier rootstock, which restricts the height of the trees. It is possible to start apple and pear trees from seed, but, eventually, without the grafting restriction, they can grow to 30 feet (9 m) or more in an ideal environment. Traditionally, crab apple is used as rootstock for apple trees, especially where they are indigenous and are hardier and more resistant to disease. The same theory applies to pears.

Choosing Your Trees
Apples

The choice to grow apples is really a personal preference—you should buy a tree that produces apples you like to eat. Do some research before you buy if you are not sure. From sweet, soft varieties to large, hard cooking apples, the choices are many. If you have the space, try a couple of different varieties. There are sweet and juicy apples, drier apples, and many crunchy, sweet apples that are also suitable for cooking, so you don't have to add as much sugar as you do with regular cooking-apple varieties.

You must also consider pollination. You may be lucky to find a self-pollinating variety, but two plants are generally required to pollinate and produce fruit. Ask for recommendations before you buy your tree so you have the right stock to produce good crops. Try and match two different varieties that will pollinate each other. Again, look for advice when you buy your tree.

Normally, apple trees will be sold as one- or two-year-old plants; it really depends on what you choose. The two-year-old tree will produce fruit a year earlier than a one-year-old tree, but there is really no other difference. Once established, an apple tree will produce fruit for decades from its third year of growth.

Pears

Your choice of pears in a supermarket is usually limited to two or three types, but there are, in fact, many different varieties that you can grow in your backyard garden. As with apples, choose the types that you and your family like to eat. Pear trees can produce fruit for more than sixty years, but the hybrid grafted trees may have shorter life spans. Some pears are self-pollinating but will produce more fruit if pollinated by another variety; check pollination advice before you buy.

Pear trees are often sold as two- or three-year-old plants, but they don't tend to fruit until they are five years old. As with apples, research growing requirements before you buy so you know what to expect.

Growing your own apples (top) and pears (bottom) allows you to explore a range of varieties not found in your supermarket.

Buying Checklist

• Consider the potential size of your tree.
• Look at planting and pruning instructions.
• Check pollination requirements.
• Make sure bark and roots are healthy and undamaged before you purchase.

All apples and pears can be bought as "cordons" that stay very small but crop heavily. Cordons are ideal for small spaces, and you can grow them very successfully against a fence or south-facing wall. Small bush varieties are good choices for slightly larger spaces, and if you have an acre or two, why not go for the full-sized versions?

Planting

Generally, apple and pear trees are not particularly picky about soil, although you should regulate extremely acidic or alkaline soil before planting to get the most from your trees. Ideally, the soil should be slightly—but only slightly—on the acidic side. Fruit trees also prefer sunny spots to give the fruit a chance to develop fully, but they will tolerate a little shade. The larger the tree, the more tolerant it will be of shade. Smaller varieties that grow in pots or against fences will need sunnier spots, and pear trees tend to need more sun than apples. The ground should be well drained, and the areas free from frost pockets or wind tunnels.

The soil doesn't have to be rich, either. Trees tend to grow well even in slightly poorer soil, although it's best to check the growing recommendations for your particular type of tree before planting. If the ground hasn't been fed or worked for a

while, a light mix of well-rotted manure or aged compost will help give the soil (and the tree) a little boost. But don't overdo the feeding; trees that live in very rich soil tend to produce a large amount of foliage but relatively little fruit.

Fruit trees should be planted from winter to spring, generally speaking. Dig over the ground about a month before planting and then dig a hole about 4 feet (1.2 m) deep and 2 feet (60 cm) wide. Incorporate any compost if you are using it. Work organic matter into the soil you dig out until it is fairly crumbly and then allow it to settle.

When you plant your tree, depth is very important. Look at the rootstock and decide where the soil line was when it was growing. This is easier to find if you bought a potted tree. With bare-root trees, the soil line isn't always obvious, but if you make sure that the "join" in the trunk—the grafted part between the rootstock and stem (or scion)—is about an inch (a few centimeters) above the ground after planting, you should be fine. Be sure not to add any more compost at this point. Simply plant your tree in the hole, fill it with the soil you previously dug out, and firm down well.

Heel around the trunk to make sure it's firmly in the ground. If the tree needs a support structure, push it into the ground when planting the tree so as not to damage the roots later. After planting, water well.

Care and Maintenance

During the first few years, make sure that your fruit trees don't dry out. When they are more established, they tend to find water by themselves. However, in long, dry periods, your fruit trees may need watering regularly to produce and develop fruit. Keep them weed free, especially when they are young. In a very cold winter, it is a good idea to mulch around the trees to protect the roots from heavy frosts. Don't mulch too close to the trunk or too high because this can cause rotting or encourage harmful insects and diseases. Remove the mulch from around the tree in the spring.

Pruning is also an important part of fruit-tree maintenance. When you plant a one-year-old tree, you generally should cut down the top part right after planting with a sharp pair of pruning shears. Make sure there are about four buds left. From the second year, prune between December and February while the tree is resting. Prune away the black shoots, leaving healthy, pink shoots to keep growing. The idea with an espalier or cordon-type apple tree or a bush pear is to train it to grow against a fence or other support.

Apple trees should have about four fruiting branches every year, but pears are better with about six branches. Once the trees are five years old, they should have established their shape and will need very little attention. However, it's a good idea to check them in the winter months so you can remove any damaged or unruly

branches. Always double-check on growing and pruning recommendations for your particular type of tree because each variety has its own special needs.

Harvesting

With larger trees, some fruit will drop naturally early in the growing season, but to get the most from your crops, you should thin them later as necessary. Look at each cluster and remove any damaged or unhealthy-looking fruit. Also, each

Use sharp pruning shears when pruning fruit trees and check each variety's special pruning needs.

Apples are ready to pick when you can twist the fruits off the branch with ease.

ripe and then ripen them off. They should be kept in an airy place out of direct light but not necessarily in the dark.

Harvest all apples and pears before there is any danger of frost and preferably before the autumn rains set in. If the weather is good and fruits are growing well, leave them on the trees until the very last minute to pick them. Use any windfall fruits as they drop from the trees rather than leaving them to rot on the ground or adding to the compost heap.

Storing

Hard fruit will keep in a fruit bowl for up to a week, and you should use softer fruits within a few days. Fruit is notoriously difficult to keep for any length of time, but harder fruits, such as apples and some of the harder varieties of pear, will store well in trays or boxes for several months. Wrap undamaged fruits individually in newspaper or tissue paper before placing them in single layers in a wooden or cardboard tray or box; store in a cool, dark place and use as needed. Too dry an environment will cause fruit to age faster, so the storage area should be a little humid but not too cold and damp.

Never store windfalls; you should use them immediately. Apples that have fallen to the ground will likely have bruising or other damage. Making fruit into pies, tarts, and preserves will add to the length of time you can store your fruit crops.

fruit needs space to develop, so removing individual fruits to give the others more room is important. It's tempting to leave them all to grow, but without the right space and air flow, fruit won't develop properly and the tree may have trouble producing any fruit at all in the following year. Cordons and some hybrids may not require any thinning, but always check on the development of the fruit and remove any that you think may be unhealthy or damaged. Damaged fruit will attract wasps and may spread further damage.

The rule of thumb when harvesting apples is to take a fruit gently in your hand and twist it slightly, and if it comes away from the tree easily, it's ready to eat. Otherwise, simply try one! You generally harvest pears before they are completely

It's also possible to dry apples and pears. To dry apples, peel and core the fruit and cut it into rings of about ¼-inch (6-cm) thickness. Soak the cut apples for about five minutes in a solution of 8 pints (4.5 liters) water and 2 ounces (50 g) salt. Drain and spread the pieces on a baking tray and dry in an oven on low heat for six to eight hours with the oven door slightly open. Cool and store in suitable containers out of direct light. You can dry apple rings in a home food dryer or on trays in the sun if you have long, sunny days in the summer.

You can dry pears in the same way as apples, but hard pears are better cut into slices and blanched for ten minutes in water and sugar rather than soaked in a saline solution. Drain and then dry.

Nutrients

"An apple a day keeps the doctor away" really does have some relevance. Apples are high in dietary fiber; they freshen your breath, clean your teeth, and exercise your gums; they are low in calories; and they are packed with vitamins and minerals. A couple of apples a day will really make a difference to your health.

The skin of apples and the flesh near the surface contain higher quantities of vitamins than the interior, and they possess properties that will help lower cholesterol and fight off certain cancers. They truly are worth adding to your daily diet.

Delicious homegrown apples and pears make it easy to add more fruit to your diet.

Apples and pears belong to the same family of plants, and pears are very similar to apples in their vitamin and mineral makeup. Apples and pears also contain pectin, which makes them ideal for making jam with soft fruits that contain less pectin.

RECIPE IDEAS

There are hundreds of recipes, both sweet and savory, for apples. Pies, cobblers, sauces, cakes, preserves, and butters are just a few suggestions. Pears are also ideal for sweet and savory dishes. Because they are often picked while still a little hard, you can soften pears by cooking them.

STEWED APPLES

Make a simple puree for a quick dessert or to accompany roast pork.

• Peel, core, and slice apples thinly.

• Put apple pieces in a pan and just cover with water. Bring to a boil and then reduce heat and simmer until soft. Harder varieties will take longer to soften.

• Drain and blend in a food processor or simply mash with a fork. Add a little sugar if desired.

APPLE PIE

• Preheat the oven to 350 degrees Fahrenheit (180 degrees Celsius).

• Peel, core, and slice 2 large cooking apples. Put in a pan and add a little water to prevent sticking. Cook over a low heat until soft.

• Drain and put apples in a lightly buttered oven-safe pie pan. Add a little water and 2 to 3 tablespoons sugar if the apples are tart. Sprinkle 1 teaspoon cinnamon on top of the apples.

• Put a pastry holder or upside-down egg cup in the middle of the dish to hold the pastry away from the apples. Roll out ready-made pie crust and lay over the top of the pie pan.

• Using a fine skewer, make a few airholes in the pie crust. Brush with beaten egg or milk and put in the oven for 30 to 40 minutes, until golden brown.

APPLES AND CHEESE

Apples are one of the best fruits to accompany cheese; here are some ideas:

• Spear cubes of hard cheese and wedges of apple onto toothpicks for a quick snack.

• Grate apple and add to bread and cheese.

• Serve apple pie with a slice of cheddar for a wonderful combination of the sweet apple and pastry and the tang of the cheddar.

POACHED PEARS

Poach pears by peeling them, removing their cores, and boiling them in a large pan of water until soft. If you want to serve the pears with a syrup, add a little sugar and vanilla flavoring to the water and then keep simmering the syrup until it is reduced after you remove the pears. Shortbread biscuits make a nice accompaniment.

PEAR AND BLUE CHEESE TARTS

Perfect for lunch or a picnic, these little tarts are very easy to make.

• Roll out four circles, about 5½ inches (14 cm) across, of ready-made puff pastry and score a border about ⅓ inch (1 cm) in from the edge.

• Mash together 4 tablespoons mascarpone and 2 ounces (50 g) soft blue cheese and then dot the mixture inside the pastry border.

• Peel and slice 2 pears and arrange the slices on top of the pastry and cheese. Glaze with beaten egg.

• Bake in a preheated oven (400 degrees Fahrenheit [200 degrees Celsius]) for 15 minutes and then scatter with a few pine nuts and a couple of sprigs of thyme. Cook for another 10 to 15 minutes until golden. Before serving, drizzle with a little honey.

OTHER IDEAS

• Add apples and pears to fruit jams and chutneys.

• Use pears in place of apples for pies, tarts, and cobblers.

• Poach pears gently in either red wine or port and serve with port for a touch of luxury.

• Slice ripe pears in half and serve with ice cream or custard.

• Pear, Stilton, endive, and walnut salad is a great winter treat. Drizzle with a little olive oil and apple cider vinegar.

Soft Fruits

Although we don't always consider soft fruits as part of our winter diet, we can enjoy that touch of summer almost all year round with a little careful planning and baking. Most soft fruits need a long and fairly sunny growing season to produce a good harvest, so a little tender loving care goes a long way, especially if you live in a less-than-tropical climate.

Strawberries

Although traditionally a summer fruit, strawberries come in a variety of hybrids that will produce fruit from early summer right through to early autumn. And you can make any excess harvest into wonderful jam to keep you going with summer taste throughout the winter months.

Strawberry plants are easy to grow and maintain and should provide you with bowls full of delicious strawberries. Wild strawberries have been growing for thousands of years, but wild plants don't produce sweet fruits. The sweeter varieties have been cultivated in home gardens since around the sixteenth century. Don't be tempted to start off your strawberry bed with wild plants unless you are planning on making a lot of wild strawberry preserves, but, even then, they will need a good amount of extra sugar to make them sweet enough to eat.

Seed

It's rare to start strawberry beds from seed, but it is possible. Buy seed from a reputable supplier and read the instructions on the seed packet before you start so you can give your plants the best possible conditions.

Generally, you should sow seed in well-drained trays or pots of fresh soil and keep them warm and watered until the plants are large enough to handle. You can then transplant them into the garden or into larger containers. Always wait until all danger of frost has passed before putting young plants outside.

Planting

There are many different varieties of strawberry available, from heavy cropping smaller fruits to larger, sweeter types. The smaller fruits are ideal if you have children to feed. There are some good, strong hybrid plants available that keep cropping into late autumn.

Gardeners generally tend to start strawberry beds with plants bought from a nursery. Or, if you know someone who has been growing strawberries for a while, you may be able to save a few of his or her plants from ending up on the compost heap at the end of the growing season. Plants put out runners, and a healthy strawberry bed needs annual thinning to keep producing healthy fruit.

You shouldn't plant strawberries in soil where you've grown peppers, potatoes, or tomatoes during the previous few years. They share the same viruses, and young plants can pick up a virus from the soil even before you get any fruit.

Find a sunny spot in your yard that you won't need for anything else during the next

Strawberries are ready for picking when they are red all over.

few years. You should renew your strawberry bed every three years or so. The ground must be well drained and, if you have organic fertilizer or well-rotted manure, dig some into the soil a month or so before planting and then lightly dig over the soil just before planting. Strawberries are ideal for raised beds. Plants tend to get straggly and are better contained if possible. Prepare the beds so you can reach the middle from both sides, making them easier to maintain.

Transplant your strawberries as soon as all danger of frost has passed, and water them well after planting. Allow at least 12 inches (30 cm) between plants and 18 inches (45 cm) between rows. Double-check the growing recommendations for your particular variety. Make sure that the roots are covered but the "crown" of the plant is above the ground.

Care and Maintenance

Keep weeds away and water your plants regularly in dry periods. When the plants have settled and are starting to grow, mulch around each one with dry straw. Don't pack it too closely; leave enough room for the plants to breathe and grow. The straw acts as a bed for your strawberries and protects them from the dampness of the soil and from slugs and snails.

When the fruits start growing, they will need protection from birds as well. Net the whole bed with a wildlife-friendly netting or, if you don't have too many birds in your

Strawberry plants can be planted in rows but are also happy in clumps or in pots and containers.

area, try hanging something to scare them, such as strings of old CDs and DVDs, over the bed to discourage them from stealing your fruit.

After the plants have finished producing fruit for the year, tidy up the bed by removing the runners. Runners are the baby plants that the mother plants send out on long, tendril-type stems. Cut the runners from the main plants with pruning shears or a sharp pair of scissors and replant them; you should do this in the early autumn or after the plants have finished fruiting. You can then leave these new plants to find their own places to take root, but the bed will become untidy and possibly overcrowded, which tends to inhibit your crops from developing well in subsequent years.

You should establish a new bed every few years, but by replanting the runners every year in a new bed, there will always be new, young plants coming up. If older plants are still cropping well, leave them for a few more seasons. When you notice that they are producing less fruit, you should remove and compost the plants and use the space for something else.

Summer-cropping strawberry varieties are vulnerable to the cold, so you should protect the plants over the winter months with a cloche or other plastic covering. Remove the cloche in early spring and agitate the soil from time to time during the winter to avoid a buildup of mold or stagnant water that can damage your plants.

Harvesting

Many growers suggest that you pick all of the fruits from your plants during the first year as soon as they appear. This is difficult when you are just getting your plants going and looking forward to a few strawberries, but the plants will put more energy into developing the roots and crown, and you will get better crops in subsequent years, so, if you can bear it, pick those first fruits before they ripen.

From the second year, pick your strawberries as soon as they are red all

over. Protect the plants from birds, even at this late stage. After picking, recover with netting and adjust the straw around the plants if necessary to stop fruits from trailing on the soil.

Storing

Most soft fruits, including strawberries, are best eaten on the day they are picked, but you can keep them in a cool place for a few days if necessary. Later-cropping varieties of strawberry will take the season right into late autumn with a little good weather, so you won't need to store them.

You can freeze strawberries, but they lose a lot of texture during the process. If you decide to freeze them, do so quickly after picking and store for a few months in the freezer. The most traditional way of storing soft fruits is in the form of preserves, jellies, and jams. With a large, heavy-bottomed saucepan and a few sterilized jars, you can

This strawberry plant has ripe fruits, ready for picking, and smaller fruits in the earlier stages of development.

make summer-fruit jams right in your kitchen and store them for many months, if not years. Homemade preserves and jams also make great gifts, saving you money and the stress of shopping.

Nutrients

Strawberries are packed with vitamins A and C and have long been used as medicinal herbs as well as delicious sweet treats. They contain a natural anti-inflammatory, which has been shown to relieve symptoms in asthma and rheumatism sufferers. The strawberry plant is also believed to relieve kidney disorders as well as throat infections, and you can use the leaves in tea infusions. And just a small handful of strawberries will count as one of your daily servings of fruit/vegetables.

RECIPE IDEAS

One of the best ways to store strawberries from your garden is to make jam. Not only will you enjoy eating it, but a jar of strawberry jam makes an excellent gift at any time of year.

STRAWBERRY JAM

To make jam, you will need a heavy-bottomed saucepan, sterilized jars, and a wooden spoon. A candy thermometer is also a handy tool but is not necessary.

• To begin, sterilize enough jars to hold about 5 pounds (2.25 kg) jam. Wash them well, rinse them in hot water, and dry them in an oven set at low heat rather than using a dish towel.

• Remove the leaves and stalks from 3 pounds (1.5 kg) strawberries and rinse gently under running water. Drain, cut into halves, and then put into a large, heavy-bottomed saucepan with the juice of half a lemon.

• Very carefully bring to a boil, stirring with a wooden spoon to prevent sticking. Reduce heat and simmer for about 25 minutes or until the fruit is very soft. Stir every couple of minutes.

• Remove from heat and stir in 3 pounds (1.5 kg) sugar. Keep stirring until the sugar has dissolved.

• Return to the heat. Bring back to a boil and boil rapidly for about 20 minutes to reach the setting point, which should read 221 degrees Fahrenheit (105 degrees Celsius) on a candy thermometer. If you are not using a candy thermometer, test for the setting point by dropping a small teaspoon of jam onto a cold saucer. After a few seconds, rub your thumb gently over the surface of the jam. Be careful—it will be hot! If it wrinkles, the jam has reached the setting point; if not, boil rapidly for a few more minutes and then test again.

• Pour or spoon the jam into warm jars. Don't put hot jam into cold jars because the glass may crack. Seal the lids and label the jars with the contents and dates.

• Store in a cool place out of direct light.

OTHER IDEAS

• Add a couple of strawberries to a smoothie. Bananas and strawberries go especially well together.

• Cut strawberries into halves or quarters and add to cereals, such as muesli.

• Turn a regular sponge cake into a strawberry delight with whipped cream and fresh berries.

• Sandwich 2 sponge cakes together and decorate the top with whipped cream and strawberry halves.

• Garnish a green salad with strawberry halves.

• The best breakfast in the world must be a few handpicked strawberries eaten while strolling around your yard or sitting outdoors early in the morning.

Blackberries

If you've ever had to deal with a bramble patch, you may not feel very kindly toward blackberry bushes. There are, however, many hybrid varieties that are thornless and will still produce juicy blackberries. Investigate different varieties at a good garden center or in seed catalogs.

People have collected blackberries from the wild for centuries, and it's tempting to just pick and eat them whenever we can. But with pollution, chemical insecticides, and other hazardous toxins floating around in the environment, you really shouldn't pick blackberries from anywhere other than clean, organic farms and gardens. In cooler climates, blackberries show up at the end of the summer and usually ripen just in time to be cooked with the first windfalls from your apple trees.

Propagation

You can tame blackberries from the wild. If you are facing a bramble patch and want to salvage some plants, arm yourself with protective clothing and a sharp pair of shears before you start. Brambles seem to have minds of their own and will happily trip you up or grab an exposed ankle with their long and very strong cable-like stems and seriously sharp thorns.

Seed

Because all plants originally start from seed of some kind, it is possible to start blackberry plants from seed. However, it will be a long time before the plants will produce fruit, and, with so many hybrid plants readily available, starting your crops from seed isn't the best method.

Layering

If you want to propagate new plants from old, lay a branch along the ground and hold it down with pegs for a few months during the autumn and winter. In ideal conditions, by the following spring, this branch should produce roots, which you then can either cut from the mother plant and replant, if necessary, or leave until the following autumn, when they will be stronger and easier to transplant. If the surface of the soil is compacted, dig a small hole to peg the branch into and cover with crumbly soil that you dug out of the hole.

Cuttings

As with most woody plants, you can start blackberry plants from cuttings. After the plants have finished fruiting in late autumn, cut 4- to 6-inch (10- to 15-cm) stems from a healthy, well-established plant. Push the cuttings, cut ends down, into well-drained pots of fresh soil. Keep them weed free and watered but don't allow the pots to become waterlogged. Transplant during the following spring if the cuttings have developed roots; however, if a hot summer is predicted, keep the cuttings in pots and transplant them during autumn instead. You can start cuttings in a seed bed, but they may need a little protection in winter.

The easiest way to start your blackberry crops is to buy plants from a reputable supplier.

If you'd rather not deal with a blackberry bush, investigate the available thornless varieties.

Blackberries can develop their root systems from the stems and can grow in many different soil types (except sandy soil).

Thornless varieties are better in smaller yards, especially if you have young children.

Planting

Blackberries are one of the few plants that will produce a heavy crop of fruit even in very shady spots. They will, however, benefit from some sun, and the newer hybrid types won't be as hardy as wild plants. Check on positioning advice when you buy blackberry canes.

Soil isn't a problem, either. The only soil in which blackberry plants don't really thrive well is dry, sandy soil. If your soil is dry and sandy, incorporate some well-rotted organic compost during the month or two before planting your canes. The ground should be fairly well drained, but, because

blackberry plants flower late in the year, mature plants are not affected by frost and won't mind a cold spot in the garden.

The usual planting time for blackberry canes is at the end of the summer or early autumn. Allow at least 18 inches (45 cm) between plants but double-check any growing recommendations for your canes if you are buying them. Some giant varieties will need several yards (meters) of space. After planting, water well. Generally, a well-established blackberry plant won't need much regular watering, but a newer hybrid may need a little more attention. Blackberry plants shouldn't be allowed to dry out because they won't produce fruit.

Care and Maintenance

Traditional blackberry plants trail, and they will crop more prolifically if supported, so train them along a fence or wall or build a support system. Blackberries, especially the thorny varieties, are useful to grow as barriers or as dividing hedges in a larger garden or in a backyard.

Blackberries produce fruit on the second-year branches. After you've picked all of the fruit, you should cut the branches that produced the fruit down to the ground and leave the other branches to produce fruit the following year.

Harvesting

Pick blackberries when they are purplish-black all over and come off the branches easily. When very ripe, the berries will drop into your hand at the slightest touch.

Blackberries don't ripen after picking, so pick only fully ripened berries. Pick regularly so that the plant develops more fruit, and so older fruit doesn't rot on the branch. The first year's growth won't produce any fruit, but you may be able to buy a plant that has second-year branches. If you plant in autumn, you should get your first crop of blackberries the following summer. Blackberries are generally ready for picking from about midsummer until late autumn, depending on your region, climate conditions, and the variety you are growing.

Storing

Blackberries will keep for a couple of days in the produce drawer of the refrigerator, but they are best eaten on the day you pick them, if possible. You can freeze the berries, but they lose a certain amount of texture and taste during the process. Freeze quickly after picking: remove any leaves, stems, and damaged fruit; place the berries on freezer trays; and put the trays in the freezer. Once the berries are frozen, put them into freezer-safe containers and label.

Blackberries will keep in the freezer for several months until you want to use them in delicious desserts.

Blackberries also make wonderful jam. Combine the berries with other fruits to produce your own special homemade preserves to give as gifts or to store for a taste of summer throughout the winter months. Apples and blackberries are a perfect combination in preserves and desserts.

Nutrients

Blackberries contain many vitamins and minerals and are particularly high in vitamin C. The berries are a source of antioxidants that fight free radicals, helping prevent certain cancers and heart disease, while the leaves have been used in alternative medicine for centuries. A cup of blackberry leaf tea every day will help boost the immune system and ward off colds and the flu as well as alleviating minor digestive disorders.

Blackberries are a wonderful treat in late summer and autumn, eaten fresh or in delicious desserts.

RECIPE IDEAS

Blackberries and apples are perfect together, and because they tend to be available at the same time of the year, it is well worth making the most of them.

APPLE AND BLACKBERRY CRISP

• Preheat the oven to 350 degrees Fahrenheit (180 degrees Celsius).

• To make the topping, simply use your fingers to combine 8 ounces (200 g) plain flour and 4 ounces (100 g) butter and then stir in 2 ounces (50 g) sugar.

• Peel, core, and slice apples and layer with blackberries in a baking dish. If you prefer, you can cook the apples for a few minutes beforehand. If the fruit is a little tangy or sharp, sprinkle a spoonful or two of sugar along with a little cinnamon over the fruit. (You can use blackberries alone if there are no apples available; taste for sweetness before adding sugar.)

• Pile the topping on top of the apples and blackberries and bake in a preheated oven for 20 to 30 minutes. Watch carefully to make sure that the topping does not burn.

• Serve warm or cold with whipped cream, ice cream, or custard.

BLACKBERRY ALASKA

This is a delicious yet simple recipe with "wow" factor.

• Make or buy a plain 9-inch (23-cm) sponge cake and put in an oven-safe dish that will double as a serving dish.

• Spread vanilla ice cream on top of the cake and level off. Cover and freeze the cake until the ice cream is firm.

• While the ice cream is refreezing, beat together three egg whites until frothy and then slowly beat in 6 ounces (175 g) sugar. Make a meringue by beating until the mixture is stiff.

• Preheat the oven to 425 degrees Fahrenheit (220 degrees Celsius).

• Remove the cake from the freezer. Prepare the blackberries by removing any leaves or damaged fruit. Place the berries on top of the ice cream. Top with the meringue mix and spread to cover the ice cream and berries completely.

• Put into the preheated oven for 5 minutes until the meringue topping is slightly browned.

• Remove and serve immediately.

BLACKBERRY PIE

• Preheat oven to 350 degrees Fahrenheit (180 degrees Celsius).

• Spoon prepared fruit into a round baking dish and sprinkle 1 or 2 teaspoons sugar over the fruit.

• Cover fruit with homemade or store-bought pie crust. Use a pastry holder to keep the crust off the fruit and therefore a little drier if desired. Brush pie crust with beaten egg or milk to glaze.

• Bake for about 30 minutes or until golden brown.

• Sprinkle a little sugar over the pie as soon it comes out of the oven. Serve with whipped cream or ice cream.

OTHER IDEAS

• If you make your own ice cream, add some blackberries to flavor it.

• Add a few blackberries to baked apples in the last ten minutes of cooking time. These are great for Halloween because the blackberry juice makes the apples look very gory!

• Use blackberries and a grated apple when making a fruitcake; stir them carefully into the mixture after you've added all other ingredients.

• Mix crushed blackberries with whipped cream when filling a sponge cake and use a few whole berries to decorate the top.

• Top ice cream desserts with a few blackberries.

• Make blackberry jam to eat in the winter as a reminder of its wonderful late-summer flavor.

Black Currants

People have collected black currants from wild plants for thousands of years. Since the Middle Ages, black currants have been used in medicinal preparations, and research in the nineteenth and twentieth centuries proved just how nutritious this fruit is.

During World War II, vitamin C was hard to come by in the United Kingdom, and the government encouraged farmers and anyone who had some land to grow black currants because this berry is probably higher in vitamin C than any other fruit grown in cooler climates. Today, black currants are one of the most widely grown commercial crops in the UK and across Europe, providing black currant juices as well as being used in many alcoholic beverages and fruit desserts.

In the United States, commercially grown black currants were banned some years ago due to a fungus from black currant plants and other plants in the *Ribus* genus transferring to pine trees and causing fatal damage to the trees. The fungus can spread over large distances—up to 1000 feet (305 m)—so the disease is hard to contain. Varieties of European and Asian pines tend to be resistant to the fungus because they evolved together. Black currants are grown in some parts of the United States, and, if you are more than 1000 feet (305 m) away from pine trees, it is well worth growing this highly nutritious and vitamin-packed fruit. Further, black currant is an excellent crop to grow in a small yard because the bushes take up relatively little space and will produce fruit for many years.

Propagation

Black currants are generally started from canes from established plants. New hybrid varieties, available from good garden suppliers, produce currants the size of grapes. Because these larger fruits tend to contain more natural sugar, less added sugar is required when cooking with these berries.

Buy plants that grow well in your region. Check with your supplier and other growers in the area, if possible, before you buy. If you do know local growers, they may be happy to donate some cuttings. Black currant bushes need pruning once a year to keep them healthy, so you may be lucky enough to find healthy canes that you can grow to start your own black currant bushes.

Cuttings

Black currant bushes are among the most successful plants to propagate from cuttings. They are hardy and don't need protection during the winter, although if your area is expecting a very bad winter, it may be better to cover them with a cloche or other type of protection during the worst of the weather. Remove the cover as soon as the weather improves; the bushes appreciate air flow and won't grow well if unventilated.

To start your black currant patch using cuttings, you need to find a healthy bush. Don't use cuttings from a diseased or a weak plant. The time to take cuttings is after you've picked all of the fruits and the leaves are falling off—usually in mid-autumn in temperate climates, although waiting a few more weeks may be better. Fruit shrubs and

Black currants are not grown widely in the Unites States, but they are a major crop across much of the United Kingdom and Europe.

Taking cuttings from healthy black currant plants is a common way of starting new plants.

trees are often planted in late autumn while the plants are at rest.

Choose a sunny spot in the yard or garden. Black currants will tolerate some shade, but berries won't develop quite as well. Although they like plenty of moisture, black currants should never be allowed to become waterlogged; also avoid frost pockets. Prepare the ground by digging a trench about 6 inches (15 cm) deep and adding well-rotted organic matter if needed. Soil should be slightly on the acidic side for the best results. If you are buying hybrid plants, check the supplier's growing recommendations before you start altering the acid/alkaline balance of your soil.

Cut 10-inch (25-cm) stems from healthy branches with plenty of buds. The cuttings should be brown, not green, and you should make your cuts just below buds. Refer to the pruning advice under Care and Maintenance.

Plant a line of canes in your trench, allowing about 8 to 10 inches (20 to 25 cm) between them. Ideally, each cane should have about four buds underground and two buds above the surface of the soil. Fill in the trench around the canes with the soil that you dug out. Try not to damage the buds above or below ground level. A little care at this point will ensure stronger, healthier plants. After planting, cut down the canes to leave just a couple of buds. This may seem drastic, but it helps develop the root system.

Water well after planting, and mulch around the canes to keep moisture in the ground. Don't mulch too close to the canes because they need some air circulation. If you don't have any mulch available, you can place pieces of old carpet or other similar material along the length of your row of canes on both sides of the plants. You can dig up the cuttings in the following autumn and move them to another position if required. Plant store-bought plants the same way as you would cuttings but, again, check for any specific growing recommendations.

Care and Maintenance

Black currants love water and should never be allowed to dry out, especially in the summer months, when they are producing fruit. Use mulch, if you have some, to keep moisture in the ground, and always remove weeds because they will draw water from the soil and take it away from your plants.

Allow the plants to grow for the first year and then, in subsequent years, prune them to keep them healthy and producing plenty of fruit. Generally, your plants need about 20 percent of the wood cut out from their middles and a little more from their outsides. Use sharp pruning shears to remove any damaged or unhealthy-looking branches as well as those that cross in the middle. Keep air circulation in mind when pruning black currants.

As long as your plants get plenty of water and air circulation and are pruned carefully once a year, they should provide plenty of luscious black currants for many years to come.

Harvesting

Pick black currants as soon as they are black. Collect them on a dry, sunny day because wet black currants rot very quickly. You can pick black currant leaves, for black currant leaf tea, throughout the year. Young leaves are preferable, but you should never take too many leaves from one plant at one time because the plant will then put more energy into producing foliage than fruit.

Storing

To store black currants for a few days, pick a cluster of berries and put them in the produce compartment of the refrigerator. They will keep for four or five days, although they are best eaten as soon as possible after picking. You can also successfully store black currant fruits in the form of pies, jams, or even wines. You can dry the leaves by hanging them in a dry, airy place and then crumbling them into a labeled jar.

Black currant canes like a sunny spot where the fruit will be exposed to full sun.

Nutrients

These unassuming little currants are packed with vitamin C, having one of the highest vitamin C levels of all fruits, while also being less acidic than citrus fruits. They have significant quantities of calcium and iron, are rich in antioxidants, and are generally a healthy crop to grow. Because of their high vitamin content, use the currants in nonalcoholic cordials and other recipes that you can bottle or freeze for the winter months.

Black currant leaf tea is popular and can alleviate some minor digestive complaints. Put a few chopped fresh or dried leaves in a jug and pour boiling water over them. Cover and allow to infuse for five minutes and then strain into a cup. You can add a spoonful of honey if you like.

RECIPE IDEAS

One of the most well-known commercial uses for black currants is black currant juice. However, if you grow your own black currants, you should try to make it yourself.

BLACK CURRANT JUICE AND ICE POPS

Kids will love these ice pops. Homemade with black currants from your own garden and packed with vitamin C, you won't need to worry about nasty additives that sometimes come with this sort of treat.

• Remove all stems and damaged fruits and put the black currants in a pan. Add sugar if you are

growing a variety that isn't sweet enough on its own, using a ratio of about 8 ounces (200 g) sugar to 1 pound (450 g) black currants.

• Cover with water and cook over low heat until the sugar is dissolved. Bring almost to a boil and then reduce the heat and simmer gently for 5 minutes or until the fruit is soft.

• Add the grated rind and juice of a lemon and continue to simmer for another 5 minutes.

• Allow the liquid to cool for 10 minutes and then strain into a warm, sterilized bottle. Seal and let cool completely. Pour the homemade black currant juice into ice-pop molds or ice-cube trays and add a lollipop stick or wooden stick to each compartment. Freeze and enjoy.

BLACK CURRANT PUREE

Mix this simple puree with other desserts, such as homemade ice cream, or use it as a filling for tarts or other pastries.

• Prepare 8 ounces (200 g) black currants and cook gently with a couple of tablespoons of water until soft.

• Push through a sieve or blend in a food processor. This will make about 5 fluid ounces (150 ml) of puree.

• To make a special creamy dessert, add the juice of half an orange when cooking the black currants as previously directed. When the puree is cool, stir in 3 ounces (75 g) powdered sugar and 10 fluid ounces (300 ml) whipped cream. Put the mixture in a freezer-safe container and freeze until firm.

OTHER IDEAS

• Use the leaves for a beneficial herbal tea.

• Use black currants in pies, either alone or mixed with other fruits.

• Black currants are especially good for tarts. Make large family-sized tarts or individual tartlets. Cook the black currants in water and sugar (if needed) and mash gently. Fill pastry shells and bake until the pastry is golden brown.

• Add power-packed vitamins to the start of your day by adding a handful of black currants to your breakfast smoothie.

• Blend black currants with a little honey in a food processor and use to top ice cream or other desserts.

• Stir black currants into a fruit salad.

• Black currants make wonderful jam. Follow the strawberry jam recipe on pages 158-159 or use a family recipe. Black currant jam made from your own homegrown berries is one of the best preserves you can have in the cupboard during the cold winter months.

RESOURCES

More Books by the Author

Grow It, Cook It (Spring Hill Books, 2011)

Granny's Book of Good Old-Fashioned Common Sense (Black and White Publishing, 2007)

Herbs & Spices (Fox Chapel Publishing's Self-Sufficiency Series, 2011)

Useful Websites

Healthy Living Books

www.healthylivingbooks.org

The author's website and blog.

Botanical

www.botanical.com

The online home of Maud Grieve's *A Modern Herbal*, published in 1931, as well as columns by gardening experts, advice on gardening equipment and supplies, and more practical guidance.

Nutrition.gov

www.nutrition.gov

Website sponsored by the United States Department of Agriculture (USDA) as a resource to help people make healthy dietary choices.

GardenAction

www.gardenaction.co.uk

A UK-based site offering a wealth of gardening articles as well as guides to growing specific fruits, vegetables, herbs, trees, and more.

INDEX

PHOTO CREDITS

ABOUT THE AUTHOR

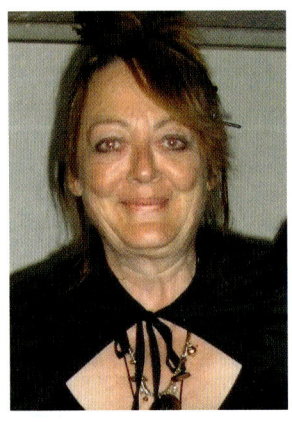

Linda Gray is a home and garden expert from London, England. She spent two years traveling with her partner and four children before settling on an acre of land in rural France. In spite of her city background, she created a garden that produced enough food to feed her family practically every day of the year. For more than ten years, she and her family were nourished with organic fruit and vegetables and enjoyed healthy lives with no need for antibiotics. Inspired to share her experiences, Linda started writing and has published more than fifty books on healthy living. She regularly writes a blog and posts articles to help readers live a happier and healthier lifestyle on her website, *www.healthylivingbooks.org*, where all of her books are also listed.